Management of Technology Systems in Garment Industry

Management of Technology Systems in Garment Industry

Gordana Colovic

WOODHEAD PUBLISHING INDIA PVT LTD

New Delhi • Cambridge • Oxford

Published by Woodhead Publishing India Pvt. Ltd.
Woodhead Publishing India Pvt. Ltd., G-2, Vardaan House, 7/28, Ansari Road
Daryaganj, New Delhi – 110002, India
www.woodheadpublishingindia.com

Woodhead Publishing Limited, Abington Hall, Granta Park, Great Abington
Cambridge CB21 6AH, UK
www.woodheadpublishing.com

First published 2011, Woodhead Publishing India Pvt. Ltd.
© Woodhead Publishing India Pvt. Ltd., 2011

Woodhead Publishing India Pvt. Ltd. ISBN 13: 978-93-80308-07-4
Woodhead Publishing India Pvt. Ltd. EAN: 9789380308074

Woodhead Publishing Ltd. ISBN 13: 978-0-85709-005-8

Typeset by SD Infosystems, New Delhi
Printed and bound by Replika Press, New Delhi

Contents

Preface

Development of Information and Communication Technologies (ICM) increasingly allows the sale and purchase of various fashion products all over the world, causing shortening the life cycle of products and reducing time of introducing products to the market. On the other hand, there comes the global competition and one can survive on the market only if all unnecessary costs are reduced, the range of production is expanded, and consumers are considered individually, not as statistical average sizes. Therefore, it is necessary to adjust production to market demands, i.e. to set a flexible production model that is capable of quick and easy adjusting to modern requirements.

Rapid technological changes and customer expectations demand from manufacturers to improve their quality of fashion products constantly and thus survive in the market. The process of making clothes is very complex and the application of the latest technological achievements is not enough for producing high-quality clothes.

Due to frequent changes in fashion trends, overcrowded markets, low purchasing power, as well as changes in habits and tastes of consumers, we are faced with a permanent decline in product sales. Changes in the world market require creating and maintaining development policy to be based on identified customer needs. Preconditions of development of workable strategies of corporate fashion industry are primarily the assessment of market potential, its own strengths and weaknesses. It is necessary to explore and explain all phenomena and laws of modern production-market-environment, in order to obtain information indicating what products to produce so that the market would accept them, and that a design as a creative discipline can create optimum products with very different characteristics.

Organization of the technological process of making clothes is different for different garments, because each item is different and requires a different organization of technological processes. Therefore, it is necessary to find the most economical ways of work and time required to perform work operations.

Production of clothing does not bring results if it does not tend to the necessity for improvements, which will lead to the growth of productivity, rational usage of productive resources and reduction of costs. It is necessary to see the growing need to change management, capacity and planning. This implies the implementation of new solutions in manufacturing, information systems, management techniques, design, etc. For successful survival in the market, it is necessary to establish control over other stages of the production cycle such as procurement, sales, promotional activities, logistics, pricing the final product, etc.

Optimization of production within the global logistic chain in the 21 century is all about the problem of determining the optimal production quantity in time, provided that the costs of purchase, costs of production, costs of storage of finished products, transportation costs and demand costs are minimal. Activities of the logistic chain begin by customer specification, and end when a satisfied buyer pays for the clothing supplied. Modern logistic chains are dynamic and flexible networks, which operate on the principle of "predict and do" versus the traditional approach of "produce and sell." Fast response to changes in demand requires solutions in all phases of the logistic chain: production, procurement, warehousing, transportation and distribution.

The world trend is to be the best, not just successful. Being competitive is not a question of success but it is the question of survival, and production business systems must be flexible, innovative and constantly improving. If the production is viewed as a chain of values that include activities which bring or do not bring the value to the product, the goal of modern production is to reduce the activities that do not bring value.

This book is the author's attempt to show, apart from introducing classical technology of production of clothing, the importance and need for improving the organization and methods of work, ways of thinking and finding new fashion markets. It is intended primarily for students of textile technology, engineers in garment industry, as well as top managers and production managers in garment industry.

I would like to thank Professor Dr Danijela Paunovic for her professional support, and Professor Sladjana Milojevic for editing.

Dr Gordana Colovic

Foreword

The author of this unique book, on the basis of years of experience and research in the field of garment industry, provides theoretical and practical examples of management and technological systems in garment industry in the region of Southeast Europe.

The dynamics of technological development goes beyond the dynamics of human perception and the difference between innovators and traditionalists brings acceptance and introduction of technology into all life processes. The path from tailor workshops to large companies goes through crises of organization. It is therefore important to organize every company adequately, according to its size, and adjust to the market economy. Clothing products are no longer the result of production but they are the products selected carefully, following the wishes of customers.

Volatility of fashion trends and modern technologies impose a permanent change in the organization of work in garment industry. The life cycle of the product is not in accordance with the life cycle of technology and it is necessary, as the author describes in Chapter 2, to define the parameters of technological systems that provide high technologics.

Flexibility and dynamics of production can be realized only through quality management. Tools for control, as well as methods for determining the time of technological operations, are described in Chapter 3 and they can be useful not only to beginners, but also to professionals experienced in this field.

To achieve the maximum level of working potential in order to increase the economy, the quantity and quality of production, it is necessary to ensure the best ergonomic conditions for workers. System, corrective, software and hardware ergonomics are shown in Chapter 4 and through ergonomic requirements they provide important factors which enable a more humane and successful work in garment industry.

Providing ergonomic principles of times, machines, production space, materials and organization a technological system can, within contemporary demands of the international fashion industry, adapt and develop business concepts in the unique world market. For customers it is not important where the product comes from but the parameters that define it through quality and price. Chapter 5 presents the analysis of planning, layout and logistics in the production of clothing as key parameters of strategic and operating management.

Modern CAD/CAM technology integrated into the CIM concept gives the advantage to producers, through the integration of all logistic activities from the moment of ordering to the delivery of finished fashion product.

Modern organizations are permanently improving, they follow the fashion changes adjusting their production capacities and adopting new methods, tools and techniques of organization of clothing production. Throughout Chapter 6 the examples of JIT concept, Toyota Production System, Kanban, PPORF and TQM system are shown, with the same aim to improve working conditions, motivate employees and increase profits. It is particularly shown in the concept of lean production and case studies.

The book is comprehensive, with numerous examples from practice, and its content is highly useful for teachers, students and those who want to enter the world of garment industry.

Dr. Danijela Paunovic

Dr. Gordana Colovic is a PhD in Industrial Management with thesis: Modeling of Flexible Garment Manufacturing (2007). She has completed her B.Sc. (Textile Technology) in 2002, and M.Sc. (Technical Science) in 2004.

She has a 22-year experience as professor with The College of Textile – Design, Technology and Management in Belgrade. She teaches modeling and construction wear and accessories, technology of garment manufacturing, organization of manufacturing, work study, management of technology systems in garment industry, and marketing management for garment industry.

She has authored 3 books and got her papers published in about 80 publications and symposiums.

1

Technology

_____

Abstract: The term technology is explained differently in different fields. In the operations management the most complete definition is that technology includes methods, means of work, production procedures, implementation by the user, as well as social relations, creative talent and sense for organization and management of knowledge in the direction of its useful application.

Keywords: Technology, organization, production.

1.1   Technology

Technology (Greek: *Tehne* = Technical skills, *logos* = science) is the application of science, a scientific method or material used to achieve the commercial or industrial aims. There are several definitions of technology:

- The United Nations Education, Social and Cultural Organisation (UNESCO, 1985) defines technology as: *"...the know-how and creative processes that may assist people to utilise tools, resources and systems to solve problems and to enhance control over the natural and manmade environment in an endeavour to improve the human condition."*
- Technology is the process of converting the value of each utility (natural or semi-finished items) to other human use values by combining business operations with the operations of machines, other mechanisms, devices, facilities, etc. which can be mechanical, chemical, thermomechanical, thermochemical, electrical, electrochemical, biochemical, etc. The entire modern industrial production is based on modern technology.
- Technology is the application of scientific or any other knowledge in organization, including any tool, technique, product, process, method, organization or system of practical tasks.

The term technology is explained differently. Most authors see it as complete definition of operation management that includes technology of management, methods, means of work, production procedures, implementation by the user, as well as social relations, creative talent and sense for organization and management of knowledge in the direction of its useful application.

Technology increasingly affects all aspects of social life. In order to survive the garment industry in the turbulent environment, it is necessary to meet customer requirements with respect to quality, price and delivery term. These criteria can be fulfilled only with the restructuring of existing production–business systems (PBS) by introducing modern technologies, changing forms of organization and participation of motivated workers. That not only changes the way goods are produced, but it also changes the manner in which the goods are distributed and promoted. Changes in technology create new markets, new products and new ways to create competitive advantage.

1.2   Cycle technologies

Adoption of technology is a common phenomenon that leads the industry through the life cycle of the industry. The life cycle of technology has seven stages:

(1) start, (2) invention, (3) development, (4) maturity, (5) uncertain, (6) slightly exhausted, and (7) obsolete.

The technology is not always accepted, even though it is a human innovation which involves generation of knowledge and processes in developing systems to solve problems and expansion of human capabilities. According to Moore theory (Moore, 1991) while accepting new technology, people can be divided into the following groups (Figure 1.1):

(1) Innovators – Enthusiasts for technology who want to be the first to test all technical innovations.
(2) Early adopters – Visionaries who are somewhat amazed at the new technology. They value the potential of products which could provide a competitive advantage for their organization.
(3) Abyss – The time gap in the acceptance of technology. It is situated between the early adopters and pragmatists.

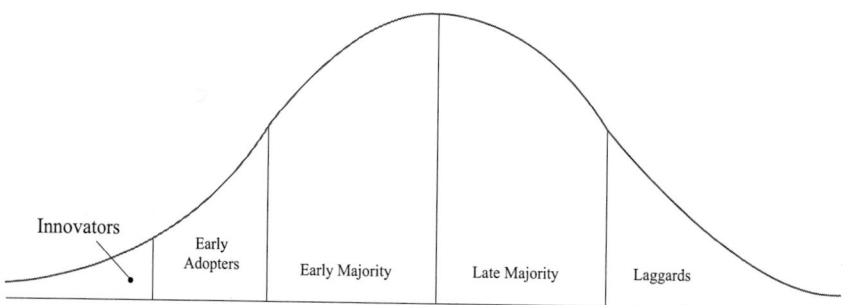

1.1 Acceptance of technology.

(4) Early majority pragmatists – People who do not like to gamble with new and innovative technologies, but are willing to give priority to technologies tested. They represent the beginning of the mass market.

(5) Late majority pragmatists – Conservatives, they represent approximately one third of the market. They do not like discontinuous innovations and believe more in tradition than in progress.

(6) Laggards – They are not interested in high-technology products.

1.3   Technology and organization

Technologies are – considering the products – different, as well as organizations. To make production technologic, every technology must require a certain way of organization.

Organization refers to functions, machines and people in workspace. Operating within a production means having good control of functions. Function refers to the man–machine interface, and organizational structure to the plant relocation and groups of people who occupy it. Organizational structure tells about the way the work is done and must be in accordance with technology.

L. Greiner (1972) developed a theory of the life cycle of the organization, which is widely accepted in the organizational literature and management practice. According to this theory, companies that grow pass through five stages of development, each of which ends with the crisis of organization: foundation and early growth, direction, delegation, co-organization and collaboration. Figures 1.2–1.6 show the stages of development from tailor workshop to large company.

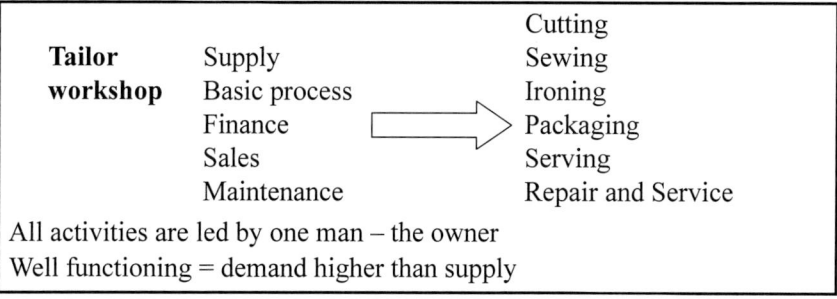

1.2   Tailor workshop.

Tailor workshop	Supply Basic process Finance Sales Maintenance	⟹	Cutting Sewing Ironing Packaging Serving Repair and Service

All activities are led by one man – the owner
Well functioning = demand higher than supply

Tailor workshop Increasing of volume	Admission of an employee The division of work – labour Main operations – other operations Informal communication Simple coordination

The main tasks are carried out by the owner, the other ones by the employee
Well functioning = further increasing of demand

1.3   Tailor shop with increase volume of work.

Further increasing of volume	Admission of new workers (ten) Group production A large number of contacts
LARGE WORKSHOP	Appointment of managers

- New division of labor –   Production
- Coordinesation ⟹ Other functions
- Training
- Specialization

Well functioning = further increasing of demand

1.4 Large workshop.

ENTERPRISE Organization	Organizational structure Technology and the study of work - Specialization - Tipization - Standardization - Preparation – production – quality - Organizational function - Centralization
Growth of demand of all products	

1.5 Production – business system.

LARGE
COMPANY
New organization
	Divided into three product lines:
	    - Women's wear
	    - Men's wear
	    - Children's wear

	Management company "from a distance" - Diversification - Control performance - Empowerment of factories - Decentralization
Profit growth Further diversification	- Autonomy - Relations among the leaders

1.6 Large company.

1.4   Technology and production

Production is the transformation of organizational resources into products. Production technology is a way of making the product, which determines the maximum amount of product from a given combination of inputs.

Classification of technology can be different, depending on the technological phase in question:

- techniques that are used in a production process,
- properties of materials,
- different skills that are required in the production process,
- degree of continuity of operations,
- degree of automation and
- degree of interdependence of business systems.

Productivity is a measure of the success of a business in relation to resources used. Productivity is not only dependent on compliance of technology and organizational structure. The relationship between technology and structure depends on the type of production, too. We distinguish the following types of production: unit production, serial production, mass production, process production and flexible production. The basic characteristics of these productions are

(1) Unit production
- single product for a known customer by order,
- requirement of high knowledge and skills of making products,
- a number of different operations without the correct order,
- implementation of universal tools and equipment,
- unpredictability of the optimal size of inventory,
- direct control of business operations,
- large number of workers of different qualifications,
- complicated linking and synchronization of operations (a source of inefficiency),
- planning and control are expensive and complex (because of uniqueness), and
- expensive and inefficient compared to other types of production.

(2) Serial production
- production of products or parts of products (series), with standardization of products and sequence of operations,
- main problem is the choice of optimal size of the series,
- fixed number and sequence of work operations,
- universal tools and equipment, grouped by type,
- large stock of raw materials located in the workshops,
- workers of different qualifications, but with a smaller range of qualifications than with unit production,
- a great need for planning and short production cycles (source efficiency) and
- delays due to waiting for the completion of the previous working operation (source of inefficiency), which is removed by the introduction of mass production.

(3) Mass production
  - production of a large number of standard products,
  - standardized capital-intensive technology,
  - uniform product of average quality,
  - specialized and line sorted equipment (conveyor belt),
  - standardized inputs, methods of operation and working,
  - narrow range of qualifications and a small number of operations per worker,
  - simple planning and control of business operations,
  - effort and flatness of work (a source of inefficiency) and
  - requirement of mass market, but changes in demand lead to combining mass produced standardized products in many variations.

(4) Production process
  - continuous production of products,
  - integrated production technology and continuous flow (processing of petroleum, chemical industry, cement production),
  - capital-intensive production,
  - mechanized and automated equipment,
  - a small number of workers,
  - the problem of planning supplies o raw materials, which is important to avoid interruption of production and
  - self-regulation and high efficiency.

(5) Flexible manufacturing
  - automated production of small series of products, without manual intervention,
  - although this technology is expensive, it provides speed, high quality, less inventory, the possibility of rapid changes, manufacturing different products and zero defect,
  - differentiated product of high quality,
  - production according to the contract (by order)
  - teamwork of multi-qualified workers,
  - more variants of basic products and
  - electronic data exchange of subcontractor brings better coordination of work

Unlike massive production, single and process productions are poorly structured and are flexible, which is achieved by a small division of labour and increased group activity, increased liability in the "role playing" and decentralization in decision-making. Flexible technological production process is an optimal model of production that enables easier and quicker adjustment to small series, to a large number of different models, different sizes and patterns, to the request of a saturated market, consumers' change of taste and to the production of different goods by using the same technological process.

According to James Thompson (James D. Thompson, 1960), technology does not unconditionally bring about the strategy of behaviour, but allows the selection of strategies for reducing uncertainty. Thompson differs:

(1) Long-linked technology – this is characterized by gradual interdependence of operations, as in mass production line. Because of the need for efficiency in this technology, the flow of operations goes according to the "Just in Time" principle, and great emphasis is given on the organization in which management controls the input and output. Long-linked technology is highly standardized and carried out in specialized serial schedule. The characteristic of this technology is moderate complexity and formalization.

(2) Mediating technology – this has a partner dependence as its characteristics, where the partners do not have to be directly dependent on each other, but are only in connection with the process of transforming input into output. Inefficiencies in this case are performed only when one side wants cooperation. The characteristic of this technology is low complexity and high formalization.

(3) Intensive technology – this technology is marked by high complexity, low formalization and high flexibility. Usual answers to different series of possibilities are given. Therefore it is not always possible to give a correct answer, due to the nature and variety of problem (e.g. in laboratory).

The efficiency of these technologies varies depending on the type of technology usable in plants. That is especially important nowadays in the situation of integrated production.

In the short term PBS has to use up the technology applied, while in the long term it can introduce a more efficient production technology in order to reduce the amount of input needed to produce a certain quantity of output.

References

1. Colovic G, Paunovic D and Savanovic G (2008). 'Buducnost evropske odevne industrije', *Tendencije razvoja u tekstilnoj industriji – Dizajn, Tehnologija, Menadzent*, DTM, Beograd, pp. 51–55.
2. Greiner L (1972). 'Evolution and revolution as organization grows', *Harvard Business Review*, July–August.
3. Levi-Jaksic M (2006). *Menadzment tehnologije i razvoja*, Cigoja, Beograd.
4. Martinovic M and Colovic G (2007). 'System PPORF in garment industry', *Serbian Journal of Management*, vol. 2, no. 1, pp. 77–85.
5. Moore GA (1991). *Crossing the Chasm: Marketing and Selling High-Tech Goods to Mainstream Customers,* New York: HarperBusiness.
6. Robbins S P and Barnwell N S (1989). *Organization Theory in Australia*, Sydney: Prentice Hall of Australia.
7. The United Nations Education, Social and Cultural Organisation (1985). *What is Technology?* Available from: http://www.pa.ash.org.au/tefa/wite.htm [Accessed 22 September 2009].

Abstract: Technological analysis involves continuous systematic testing
of alternative permutations of production and changes of technological
operations and a synthesis of future technological processes. Organization
of the technological process of sewing and finishing is different for
different garments. Each product is different in its own way and requires
a different organization of the technological process of sewing and
finishing. Well-selected technological operations shorten the time of
making garment cases, reduce production costs per unit of product,
allowing the flow of product through all stages without the occurrence of
bottleneck production, reduce inventory, allow rational use of the machine
park and prevent low labour productivity.

Keywords: processes, technological analysis, manufacturing operations,
garment.

2.1   Technological system

Every production, every organized human labour is a complex system.
Technological system is an open dynamic system closely related to the
environment. Production technological system is designated as a part of a
broader production system (element of the business system). Business system,
in organizational terms, can act as a separate entity (company). Business
system, beside production system, also contains a system of procurement,
sales, distribution of resources, as well as material, energetic, informational
and financial flows. Socio-economic system is broader than business system
(Figure 2.1).

The basis of technological system is in the process, transforming materials
from one form into another, from lower to higher use value, which directly
determines the character of the production system (Figure 2.2). Other parts of
the production system are

- System of design (construction) of product,
- System maintenance,
- Inventory system,

- Safety at work,
- Transport,
- Quality control.

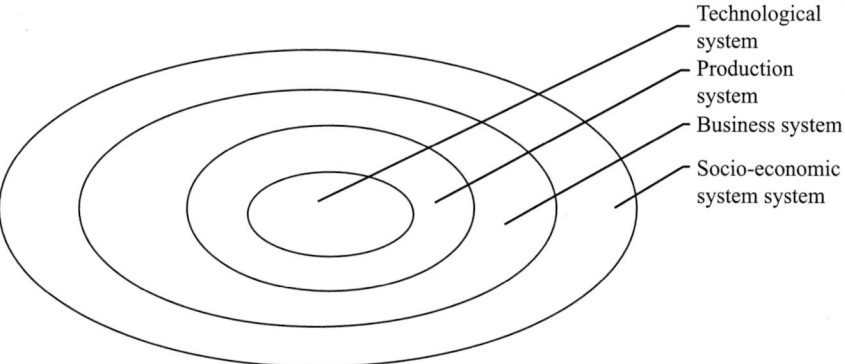

2.1 Systems.

Fiber ⟹ Yarn ⟹ Fabric, knitted ⟹ Clothes

2.2 Transforming materials from one form into another.

2.2   Technological systems, processes and operations

Technological system usually occurs as a part of a wider system and the result of an integral activity of people in different kinds of work processes.

The structure of the technological system is determined by three factors:

(1) Complexity of technology,
(2) Complexity of products and
(3) Management system.

Technological systems by nature are among the artificial, open, dynamic and stochastic systems. Technological systems are studied both in the sphere of production and beyond, so they are mainly divided into production and non-production technological systems.

Production technological systems can be defined as a set of objects (tools, materials, funds for the work, projected technology, human labour and finished products) with the relations that exist between input elements on one side and output elements (finished products) on the other, observed through their attributes (price, quantity and quality). Non-production technological systems occur in all out-production activities of people (education, health, culture, etc.).

The essence of the production technological system is a mutual dependence and interdependence of all elements (or objects of system) while performing the functions of transformation of material from one form into another, more useful form, where its utility output increase under the influence of organized human labour. Figure 2.3 shows the technological process of making shirts for men.

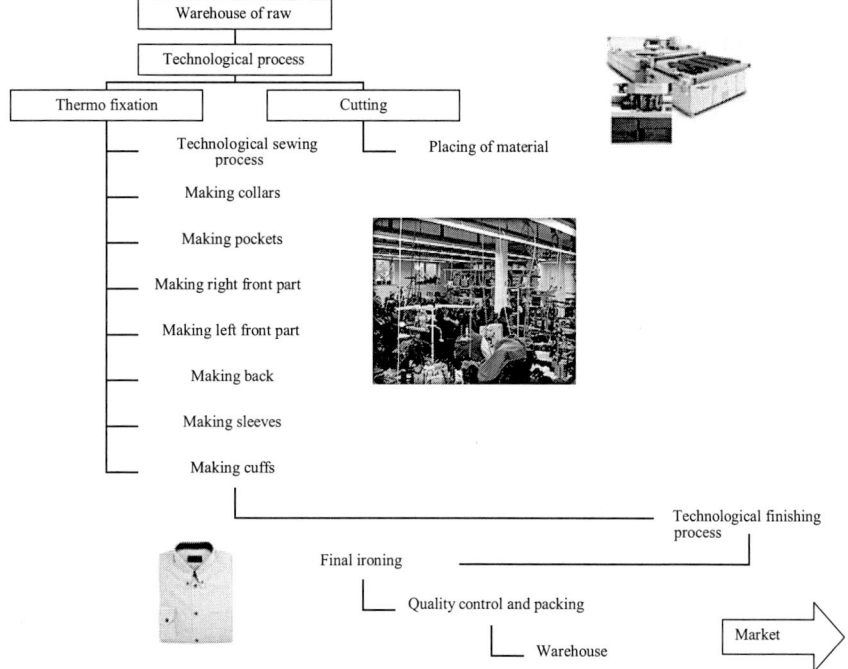

2.3 The technological process of making men's shirts.

Production technological systems are classified according to the following:

(1) Level of investment (of raw materials and simple compounds, drawer, basic compounds, sub-assemblies and complex materials, components and final products),

(2) Type of labour (extractive, processing and synthetic technological processes),

(3) Type of labour and types of activities (agricultural, mining, metallurgical, chemical, metal-processing, textile, pharmaceutical, wood and food),

(4) Dynamics of movement of materials and stability conditions (batch wise or continuous),

(5) Organization of production (mass, serial and unit production),

(6) Order of processes (preparation of raw materials, chemical processing, physical processing and finishing) and

(7)   Other criteria (the character of the means of work, production volume, product type, the basic raw materials and the dynamics and type of movement of material in the technological process).

Processes in production are a horizontal division of labour whose task is to make the product. Production process includes everything that happens with the subject from the entry of raw materials in production to the release of finished products. The production process consists of elementary processes: workplaces, quality control, inter phase transport, preventive maintenance of the means of work, preventive work safety, storage and supply of water and energy.

Technological process is part of the production process which refers to the shaping of work case with defined workplaces. Technological process is the linking of technological operations in order to convert the lower use-values into the higher ones together with human activity. Technological operation is a set of direct and ancillary effects on the work piece on one machine, which enables the realization of process. Working operation is a set of all activities that form a finished product.

Operations can be divided into technological and non-technological. Technological operations directly alter the characteristics of objects to get products with new use-value on the basis of these changes. Non-technological operations do not change the characteristics of objects, but are necessary in the production process so the technological process could be done.

2.3   Technological analysis of manufacturing operations

Technological analysis involves continuous systematic testing of alternative permutations of production and changes of technological operations and a synthesis of future technological processes. Optimization of technological system means the ultimate goal of the analysis of technological systems and is an element of his partial analysis.

The main objective of the analysis of technological systems is to improve performance through analysis of process. Technological analysis determines the effect of technological change in operations on the broader changes of technological system, as well as on the performance of certain operations.

Changes in the technological operations are viewed via

- fixed costs of working capital,
- expenditure of human labour,
- appropriate changes in the course and the amount of material and
- changes in all other operations of technological process.

Technological analysis is a specific activity, which aims to introduce the production characteristics of products and potential problems that we are about to have in its production. The greatest number of errors in garment manufacturing, and thus the costs associated with product quality arise in defining garment, developing product and planning of technological process of making clothes. It is believed that 75% of all errors that appear on the product occur in construction preparation. The most common errors in construction preparation are

- pattern pieces do not fit the model,
- bad positioning of pattern pieces,
- unmarked indentation,
- missing of pattern piece,
- adding % due to stretch material,
- deviations in grading,
- ill-cut pattern pieces,
- non-grading of all pattern pieces,
- large consumption of materials,
- pattern pieces not fitting the layout pattern and
- inadequate size of layout pattern.

The manufacturing process is a database of functioning of organizational structures, which requires being technologic. The technologics of product is achieved through such construction of product that ensures an optimal relationship between investment of resources and the achieved quality under the given driving conditions and the absorbing power of markets.

Therefore, it is necessary for the design of technological products to undergo technological analysis, in order to determine and, if it is necessary, to improve the technologics of product, i.e. the suitability for production. It is necessary to observe the possibilities of one's own production facilities. Figure 2.4 show the functional clothing design system that provides high technologics.

While planning of production of each garment a detailed technological analysis needs to be made. The technological preparation consists of analyzing, enhancing and improving of activities related to technological processes, which can be divided into several groups of activities, such as technological analysis of production operations, the selection of machine, montage plans, selection of technological systems, the choice of inter phase transport system, the choice of mounting positions, determining the technological and technical specifications for the programming of machines, work study and workplace design.

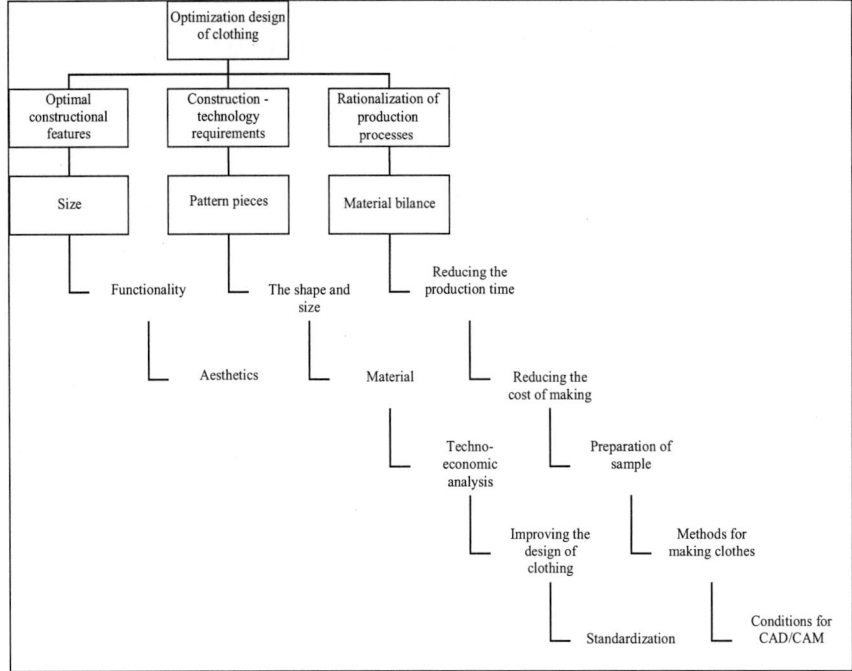

2.4 Functional clothing construction systems that provides high technologics.

Modern fashion design requires a small amount of clothing, many colours and patterns, so the production plants daily deal with many work orders, which caused the production of technical documentation to be one of the biggest problems in clothing industry.

Organization of the technological process of sewing and finishing is different for different garments; for each item is different in its own way and requires a different organization of the technological process of sewing. Well-selected technological operations shorten the time of making garment cases, reduce production costs per unit of product, allowing the flow of product through all stages without the occurrence of bottleneck production, reduce inventory, allowing rational use of the machine park, preventing low labour productivity and so on. Therefore, the task of technical preparations is to determine working procedure for the new product, to determine the required time of manufacture, the material normative, and to match the way of making with some details. On the basis of daily capacity, the required number of workplaces should be determined, as well as the number of ordinary and special sewing machines, automatic sewing machines and presses for trim, tables and other tools of work, the number of workers in structure with highly specified load job.

In garment industry, technological process is divided into three phases: cutting, sewing and finishing. Each phase individually requires plans of technological operations. A plan of technological operation (operation sheet) is the basic document in the development of a garment, on the basis of which other technological documentation is made.

After making an operation sheet the recapitulation of a development time is performed, according to the types of machines used for making a garment and time required for manual work to determine the number of necessary funds. Total production time per unit (t_1) is obtained by adding the time of making, following the stages of production:

$$t_1 = t_c + t_s + t_f \qquad [2.1]$$

Where t_c – cutting time,
t_s – time for sewing phase,
t_f – time for finishing phase.

Making plans and technological processes is a complex and responsible job which requires integration of knowledge in order to achieve the optimization of process parameters of production of clothing. Due to the lack of time and professional staff in the garment industry, less technological documentation is rarely made or used. Steady production lines for the production of certain garments are often used, regardless of the size of work orders.

2.3.1 Technological analysis of operations for making men's shirts from denim

Analysis of technological operations in the process of cutting and sewing men's shirts from denim is given as an example of technological analysis (Figure 2.5). Total production time per unit (t_1) is 3336 s. Time of cutting shirts for men from denim (jeans) is 295 s and time for sewing and finishing phase is 3041 s (Table 2.2). Table 2.1 shows the need for three workers for cutting because total load is 300%.

Table 2.1 Technological operation plan for cutting men's shirts from denim

Name of operation	Means of production	Pr quota/a piece (s)	Norma (piece)	Load (%)
Marking length of cutting layout (marker) after patterns and spreading of material (cutting layers)	Fabric Spreading machine	21	1258	21

Putting of cutting layout on material	Hand makes	4	6600	4
Rough cutting	Straight knife Cutting machine	31	851	32
Fine cutting	Vertical cutter	59	447	60
Numbering, marking of cut pieces	Hand makes	49	538	50
Completing of cut pieces	Hand makes	53	498	54
Control	Hand makes	78	338	79
TOTAL TIME		295 s		

Table 2.2 Technological operation plan for making men's shirts from denim

Name of operation	Means of production	Pr quota/a piece (s)	Norma (piece)	Load (%)
Open bundle and control of cutting pieces	HR	33	800	29
Preparation for sewing collars	HR	16	1520	15
Making collars	OM	40	660	36
Turning and shaping collar tops	RR	54	2020	52
Topstitch collar	SM2	40	650	36
Cut the tops of collars	HR	10	2500	9
Hem stand collars	OM	22	1200	19

(Continued)

Hem cuff and sewing	OM	21	1230	19
Hem two pockets	OM	14	1800	13

Prepare the cover for pocket for sewing	HR	30	905	26
Sewing covers for pockets	OM	77	340	69

Turning the covers for pockets	HR	42	600	39
Topstitch covers for pockets	SM2	122	210	112

Sewing stand collar on collar	OM	66	400	59

Cutting and turning the stand collar	HR	23	1100	21
Topstitch the stand collar	OM	37	700	34

Sewing yoke on the front part	OM	53	500	47

| Topstitch the yoke | SM | 42 | 610 | 38 |

| Pressing the front parts | HR | 35 | 750 | 31 |
| Closing placket | OM | 89 | 300 | 78 |

Ironing pockets	SI	20	1300	18
Mark with pattern place for sewing pockets	HR	37	713	33
Sewing pockets	OM	220	120	196

| Sewing covers for pockets | OM | 52 | 507 | 46 |

| Sewing yoke on back | OM | 20 | 750 | 31 |

(Continued)

Topstitch the yoke	SM2	80	300	78
Sewing labels on the yoke	OM	13	1886	12
Sewing shoulders	SMo5	37	700	34
Topstitch the shoulders	SM2	52	507	46
Facing the two sides of the sleeves fly	OM	66	400	59

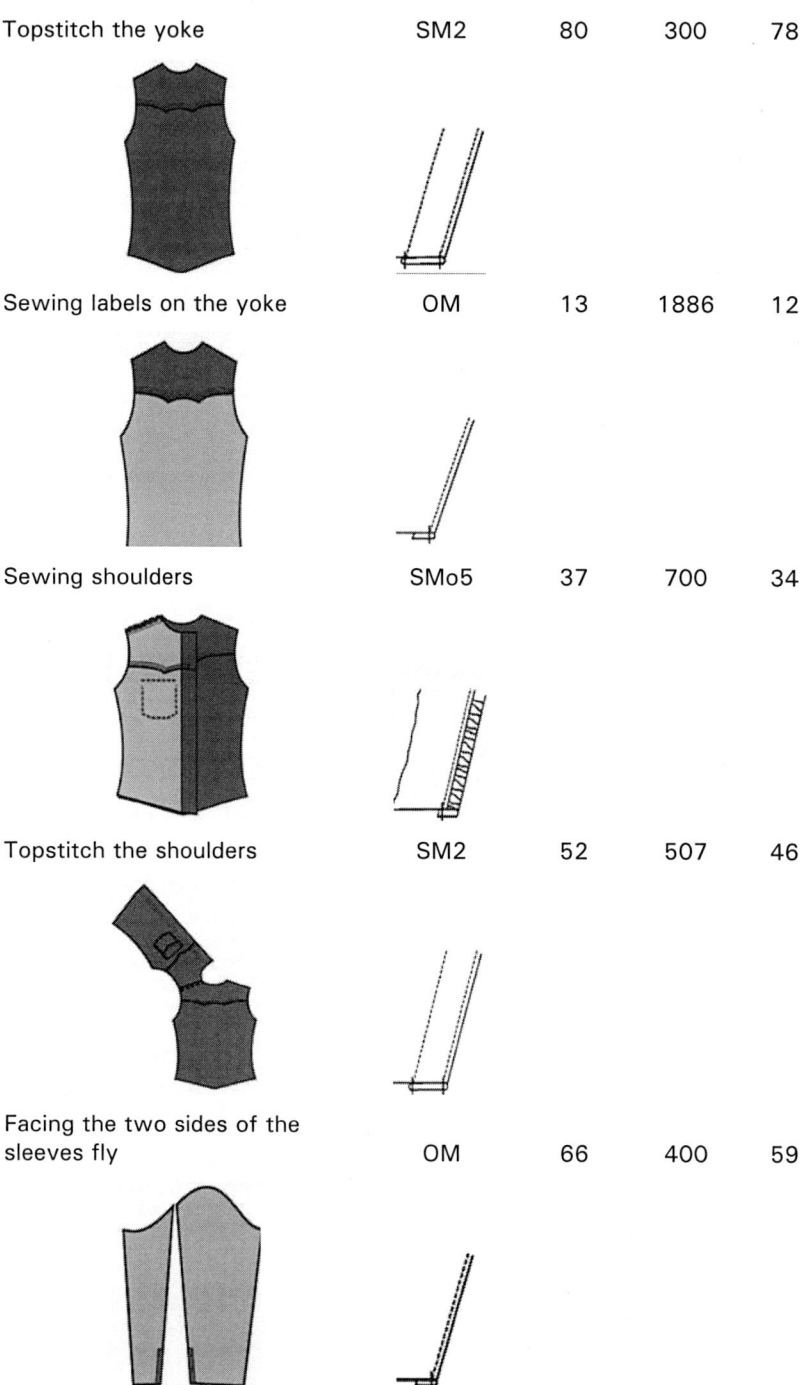

Sewing sleeves	SMo5	38	694	34

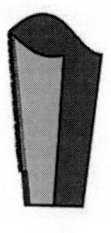
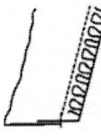

Topstitch the sleeves	SM2	35	754	31

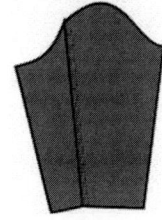

Attaching sleeves to armholes	SMo5	88	300	78

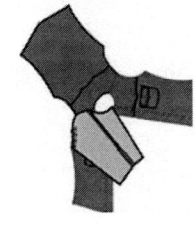
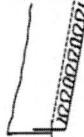

Topstitch the armholes	SM2	132	200	118

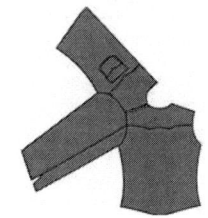

Sewing underarms seams and side	SMo5	88	300	78

(Continued)

Making cuffs	OM	48	550	43
Turning and shaping cuffs tops	HR	43	610	38
Attaching cuffs on the sleeves	OM	97	270	87

Close the cuffs on the sleeves	OM	66	400	59

Topstitch the cuffs	OM	66	400	59

Attaching collar to neckline	OM	66	400	59

Close collar	OM	66	400	59
Hem shirt	OM	52	500	47

Sewing ten buttonholes	AUTh	88	300	51

Close shirts at the front (protection when sanding)	OM	52	500	30
Wearing sleeveless shirts (protection of the stoning)	HR	52	500	30
Cutting of thread	HR	264	100	153
Ironing on the shirt ironing machines	Shirt Finisher	132	200	76
Final ironing (cuffs, collar)	SI	32	810	19
Sewing buttons	AUTb	26	1000	15
Putting paper labels	HR	26	100	153
Packaging shirts in the bag	HR	17	1553	10
TOTAL TIME		3037s		

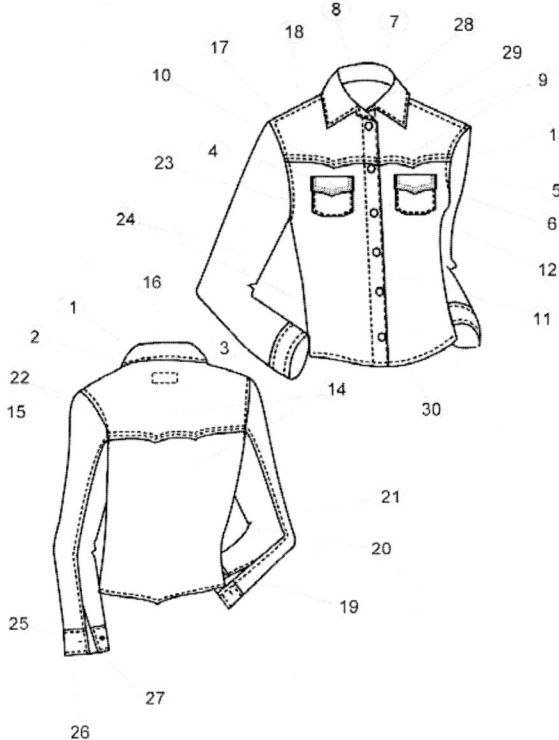

2.5 Men's shirts from denim.

Required number of workers in the process of cutting is provided in the following way:

$$Nw = C_d \times t_1/T = 2.8 \approx 3 \text{ workers} \qquad [2.2]$$

Where, C_d – daily capacity,

$\quad$ t_1 – total production time per unit,

$\quad$ T – working time.

Production line with 27 workers produces (daily capacity) 247 pieces of shirts from denim per day. In the plan of technological operations (Table 2.2) the time of making this operation is given, together with the norm (number of pieces you need to do for the working time 7.5 h), the load of work operations in percentage, and the necessary means of work. Means of work are marked with the following abbreviations:

- Hand makes – HR
- Ordinary sewing machine – OM
- Special sewing machine with two needles – SM2
- Special sewing machines (overloch) with three threads – SMo3
- Special sewing machines (overloch) with five threads – SMo5
- Automatic for making buttonhole – AUTh
- Automatic for button – AUTb
- Steam iron – SI

2.3.2 Technological analysis of operations for making women's shirts

Cutting time for women's shirts is 295 s (three workers). Technological operation plan for cutting women's shirts is shown in Table 2.3. Production lines with 29 workers produce 332 shirts per day. Model of women's shirts with positions of some of the operations are shown in Figure 2.6.

Table 2.3 Technological operation plan for cutting women's shirts

Name of operation	Means of production	Pr quota/a piece (s)	Norma (piece)	Load (%)
Spreading material	Hand makes	36	750	44.3
Spreading nonwoven interlining	Hand makes	13	2077	16.0
Rough cutting (without front parts)	Straight knife Cutting machine	11	2455	13.5
Fine cutting (with front parts)	Vertical cutter	28	964	34.4

Numbering and marking of cut pieces	Hand makes	22	1227	27.1
Fusing interlining with collar	Fusing machine	34	794	41.8
Thermal bonding interlining with cuff	Fusing machine	45	600	55.3
Completing of cut pieces	Hand makes	25	794	41.8
Control	Hand makes	30	900	36.9
TOTAL TIME		244 s		

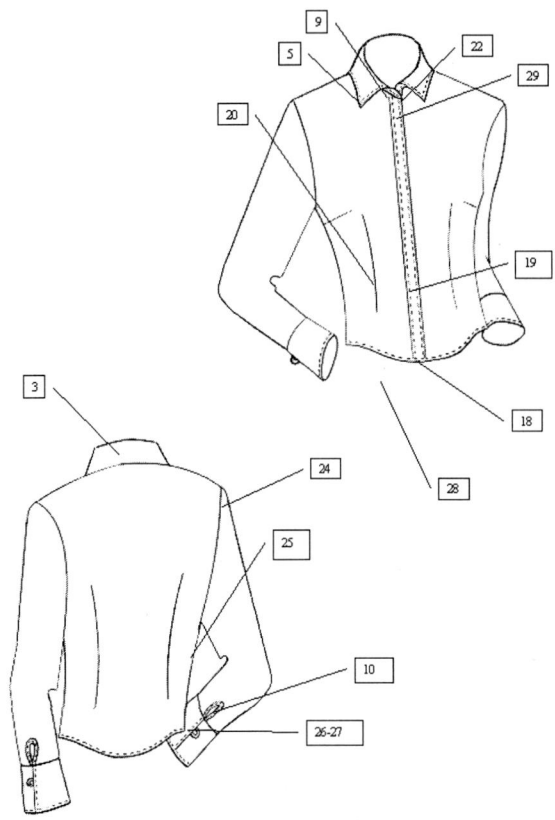

2.6 Women's shirt.

Technological operation plan for the production of women's shirts is shown in the Table 2.4.

Table 2.4 Technological operation plan for making women's shirts

Name of operation	Means of production	Pr quota/ a piece (s)	Norma (piece)	Load (%)
Making collars	OM	54	500	66.4
Turning and shaping collar tops	HM	40	675	49.2
Ironing collar	SI	62	435	76.3
Topstitch the collar (0,5cm)	OM	43	628	52.9
Hem stand collar	OM	21	1286	25.8
Sewing stand collar on collar	OM	54	500	66.4
Turning and shaping stand collar tops	HM	40	675	49.2
Topstitch the stand collar	OM	18	1500	22.1
Placing strips on the sleeve	SM	58	466	71.2
Hem cuffs	OM	39	692	48.0
Sewing cuffs with long side	OM	10	2700	12.3
Ironing cuffs with long side	SI	45	600	55.3
Making loop for button	SM	59	458	72.5
Sewing cuffs with short side with putting loop for button	OM	53	509	65.2

Turning cuffs	HM	40	675	49.2
Ironing placket	SI	75	360	92.2
Sewing placket on front parts	OM	40	675	49.2

| Topstitch the placket for 0,5cm | OM | 65 | 415 | 80.0 |

| Sewing darts on front parts and back and sewing bust darts | OM | 80 | 338 | 98.2 |

| Sewing shoulders | SMo5 | 27 | 1000 | 33.2 |
| Attaching stand collar to neckline | OM | 145 | 186 | 178.5 |

| Closing stand collar | OM | 61 | 443 | 74.9 |
| Attaching sleeves | SMo5 | 54 | 500 | 66.4 |

(Continued)

Sewing side seams and sleeves with label	SMo5	45	600	55.3

Attaching cuffs	OM	125	216	153.7
Topstitch the cuffs for 0,5cm	OM	133	203	163.5

Hem	OM	93	290	114.5

Marking and sewing seven buttonholes	AUTh	75	360	92.2

Marked place for button	HM	61	443	74.9
Sewing seven button	AUTb	98	276	120.3
Ironing	Finisher	309	87	381.6
Control	HM	140	193	172.0
Mount the hanger	HM	10	2700	12.3
Buttoning	HM	62	435	76.3
Putting paper labels	HM	10	2700	12.3
TOTAL TIME		2344s		

The technological documentation shows that the total production time per unit (t_1) is 2588 s.

2.3.3 Technological analysis of operations for making women's denim jacket

Production line with 47 workers produces 293 women's denim jackets per day (Figure 2.7). Cutting time is 369 s (four workers). Technological operation plan for the cutting women's denim jacket is shown in Table 2.5.

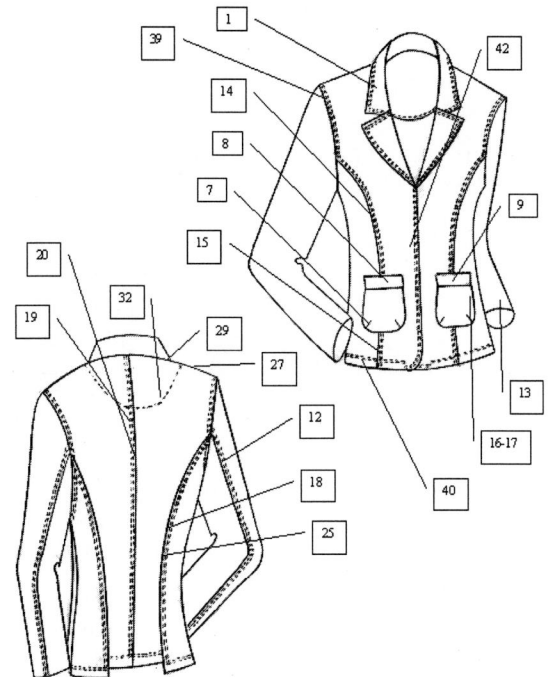

2.7 Women's denim jacket.

Table 2.5 Technological operation plan for cutting women's denim jacket

Name of operation	Means of production	Pr quota/ a piece (s)	Norma (piece)	Load (%)
Marking length of cutting layout after patterns; spreading matherial	Hand makes	37	730	40.2
Planing of cutting layout on matherial	Hand makes	15	1800	16.3
Rough cutting	Straight Knife Cutting Machine	55	491	59.7
Fine cutting	Vertical Cutter	31	871	33.6
Sreading elastic bar	Hand makes	15	1800	16.3
Rough cutting elastic bar	Vertical Cutter	5	5400	5.43

(Continued)

Numbering and marking of cut pieces	Hand makes	129	209	140
Completing of cut pieces	Hand makes	35	771	38
Control	Hand makes	47	574	51
TOTAL TIME		369 s		

Technological operation plan for the production of women's denim jacket is shown in Table 2.6.

Table 2.6 Technological operation plan for the production of women's denim jacket

Name of operation	Means of production	Pr quota/a piece (s)	Norma (piece)	Load (%)
Sewing collar	OM	65	415	70.6

Name of operation	Means of production	Pr quota/a piece (s)	Norma (piece)	Load (%)
Cutting and turning collar	HM	40	675	43.4
Ironing middle collar seam	SI	25	1080	27.1
Hem front fuse and double bend	OM	50	540	54.3
Ironing fuse behind the neck	SI	21	1286	22.8
Sewing hangers for fuse	OM	8	3375	8.68
Sewing dart on pockets	OM	60	450	65.1

Name of operation	Means of production	Pr quota/a piece (s)	Norma (piece)	Load (%)
Sewing elastic bar on pockets	SMo5	21	1286	22.8

Name of operation	Means of production	Pr quota/a piece (s)	Norma (piece)	Load (%)
Topstitch the pockets	OM	62	435	67.4

Name of operation	Means of production	Pr quota/a piece (s)	Norma (piece)	Load (%)
Sewing sleeves	SMo5	59	458	64
Topstitch the sleeves	OM	145	186	158
Sewing side seam on sleeves	SMo5	59	458	64

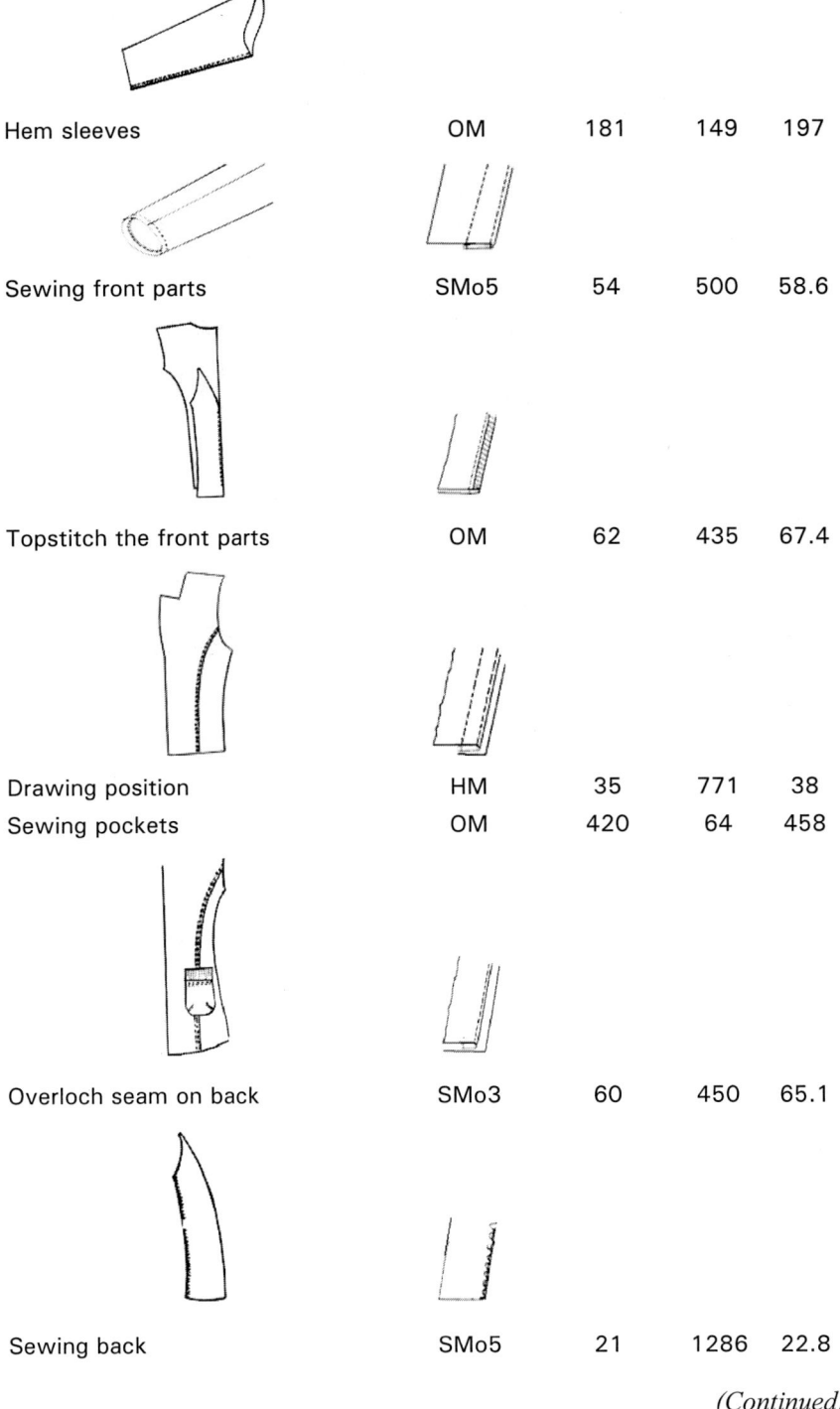

Hem sleeves	OM	181	149	197
Sewing front parts	SMo5	54	500	58.6
Topstitch the front parts	OM	62	435	67.4
Drawing position	HM	35	771	38
Sewing pockets	OM	420	64	458
Overloch seam on back	SMo3	60	450	65.1
Sewing back	SMo5	21	1286	22.8

(Continued)

Topstitch the back	OM	39	692	42.3

Sewing back with side parts to fly	OM	43	628	46.7

Closing fly on the back with formation hem	OM	92	293	100
Topstitch the back hem	OM	54	500	58.6
Closing fly on the side to hem	OM	92	293	100
Sewing shoulders	SMo5	27	1000	29.3
Topstitch the shoulders on back	OM	77	351	83.5

Attaching collar	OM	63	429	68.3

Seaming front fuse with back fuse	OM	9	3000	9.77

Operation	Machine	Col1	Col2	Col3
Sewing fuse on back	OM	51	529	55.4
Topstitch the fuse	OM	91	297	98.7
Topstitch the collar, lapel to end front fuse on hem	OM	312	87	337
Sewing the side seam with label	SMo5	54	500	58.6
Topstitch the back	OM	62	435	67.4
Attacking sleeves	OM	185	146	201
Placing strips on the armholes	OM	497	54	543
Sewing armholes	OM	298	91	322
Hem	OM	195	138	212
Sewing two buttonholes	AUTh	30	900	32.6
Sewing two button	AUTb	16	1688	17.4
Cutting of thread	HM	180	150	195

(Continued)

Ironing	HM	245	110	266
Control	HM	145	186	157
Mount the hanger	HM	6	4500	6.51
Buttoning	HM	9	3000	9.77
Putting paper labels	HM	8	3375	8.68
TOTAL TIME		4263 s		

2.3.4 Technological analysis of operations for making women's trousers

Production line with 28 workers produces 294 pieces of women's trousers per day (Figure 2.8). Time needed for cutting is 347 s (three employees). Technological operation plan for the cutting women's trousers is shown in Table 2.7.

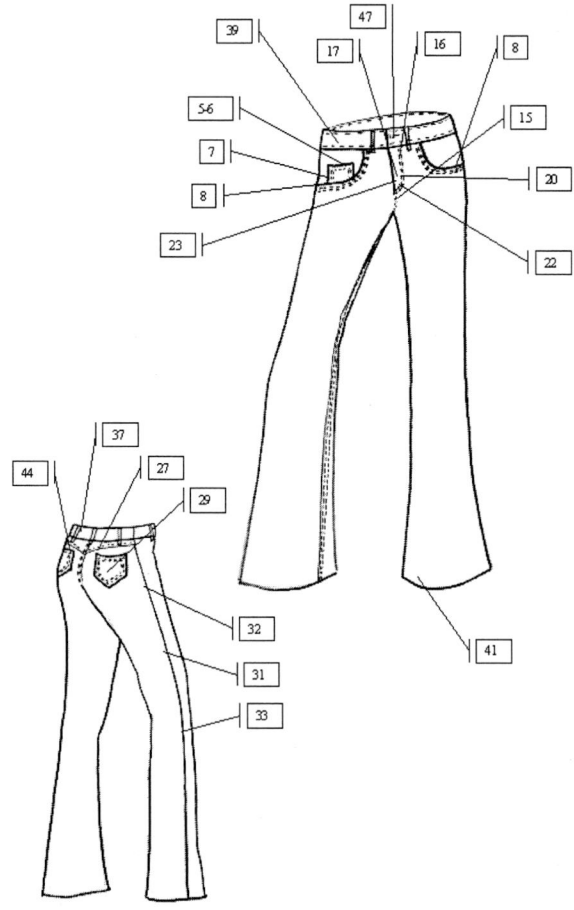

2.8 Women's trousers.

Table 2.7   Technological operation plan for the cutting of women's trousers

Name of operation	Means of production	Pr quota/a piece (s)	Norma (piece)	Load (%)
Marking length of cutting layout after patterns; spreading material	Hand makes	45	600	39
Planing of cutting layout on material	Hand makes	17	1588	14.74
Rough cutting front and back legs	Straight knife Cutting machine	47	574	40.77
Fine cutting	Vertical cutter	24	1125	20.8
Marking length of cutting layout for interlining for pocket ; spreading interlining for pocket	Hand makes	17	1588	14.74
Fine cutting interlining for pocket	Vertical cutter	7	3857	6.067
Numbering, marking of cut pieces	Hand makes	116	233	100.4
Completing of cut pieces	Hand makes	32	844	27.73
Control	Hand makes	42	643	36.39
TOTAL TIME		347 s		

Technological operation plan for the production of women's trousers is shown in Table 2.8.

Table 2.8   Technological operation plan for the production of women's trousers from denim

Name of operation	Means of production	Pr quota/a piece (s)	Norma (piece)	Load (%)
Hem watch pocket	OM	10	2520	9.3
Ironing watch pocket	SI	20	1260	18.6
Making position for watch pocket	HM	10	2520	9.3
Sewing watch pocket	SM2	52	485	48.2
Sewing in-pocket on pocket bag	OM	45	560	41.8

(Continued)

Sewing lacket on pocket bag	OM	53	475	49.3
Sewing pocket bag	SMo5	40	630	37.1

Topstitch hole of pocket	SM2	133	189	123.8

Closing pocket bag	SMo5	44	573	40.8
Turning pocket bag	HM	6	4200	56
Topstitch pocket bag	OM	23	1096	21.4
Sewing pocket bag with front part (leg)	OM	43	586	39.9

Overloch seam on fly	SMo3	12	2100	11.1
Turning on half and overloch seam on underlap	SMo3	10	2520	9.3
Overloch seam front parts in part of underlap	SMo3	30	840	27.9

Sewing zipper on fly (3,5cm and 3cm)	OM	38	663	35.3

Sewing fly on left front part	OM	50	504	46.4

Topstitch on part of fly	OM	60	420	55.7
Making position for topstitch on fly	HM	10	2520	9.3
Topstitch with part of fly	SM2	33	764	30.6

Sewing underlap with zipper	OM	30	840	27.9
Sewing right front part on underlap and sewing part under fly	OM	60	420	55.7

Closing front part under fly	OM	45	560	41.8

Hem on back pocket	OM	55	458	51.1
Market position for embroidery on back pocket	HM	45	560	41.8
Embroidery on back pocket	OM	95	265	88.3
Ironing back pocket	SI	65	388	60.3
Sewing yoke with back part	SMs	60	420	55.7

Market position for back pocket	HM	60	420	55.7

(Continued)

Sewing back pocket	OM	210	120	195
Seat seaming	SMs	65	388	60.3
Overloch side seam	SMo3	60	420	55.7
Sewing side seam	OM	90	280	83.6
Topstitch side seam on back part to the end of pocket bag	OM	40	630	37.1
Inseam legs	SMs	180	140	167.1
Making belt loops	SMbl	15	1680	13.9
Cutting belt loops (5cm x15cm)	HM	15	1680	13.9
Sewing belt loops and label	OM	42	600	39
Making belt	SMb	70	360	65

Closing belt	OM	80	315	74.3
Hem legs	OM	120	210	111.4

Making bartack on fly (one)	SMbt	8	3150	7.4
Making bartack on underlap (one)	SMbt	8	3150	7.4
Making bartacks on back pocket (four)	SMbt	20	1260	18.6

Making bartacks on belt loops (ten)	SMbt	75	336	69.6
Making bartacks on belt	SMbt	15	1680	13.9

Making rivets on watch pocket and hole of front pocket (six)	PPr	83	304	77
Making metalic button on belt	PPr	7	3600	6.5
Cutting of thread	HM	300	84	278.6
Ironing	Finisher	305	83	281.9
Control	HM	140	180	130
Closed pants	HM	3	8400	2.8
Putting paper labels	HM	10	2520	9.3
TOTAL TIME		3198 s		

A special sewing machine for closed seams (SMs) can be used for the production of a special machine for making belt loops (SMbl), a special machine for making belt (SMb), machine for bartack (SMbt), and pneumatic presses for rivets (PPr).

2.3.5 Technological analysis of operations for making sweat

Production line with 10 workers produces 185 pieces of sweat per day (Figure 2.9). Cutting time for three workers is 443 s. Technological operation plan for the cutting of sweat is shown in Table 2.9.

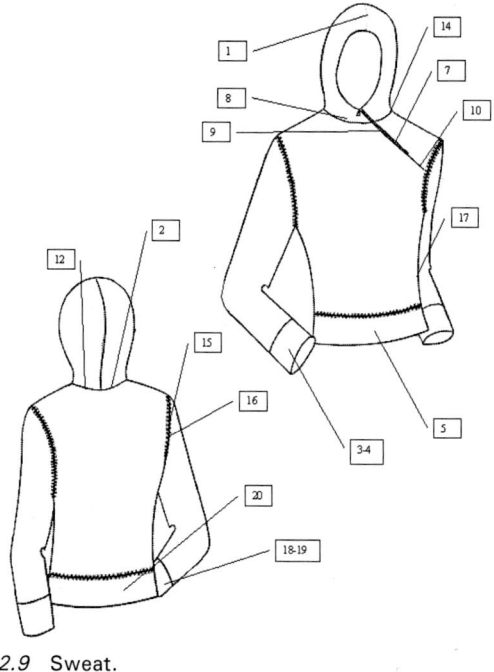

2.9 Sweat.

Table 2.9 Technological operation plan for cutting sweat

Name of operation	Means of production	Pr quota/a piece (s)	Norma (piece)	Load (%)
Marking length of cutting layout after patterns; spreading matherial	Hand makes	22	1227	15.1
Planing of cutting layout on matherial	Hand makes	4	6750	2.7

Rough cutting: front part, back, sleeves, hood, belt, cuff	Straight knife Cutting machine	32	844	21.9
Fine cutting: front part, back, sleeves, hood, belt, cuff	Vertical cutter	59	458	40.4
Numbering, marking of cut pieces	Hand makes	196	138	134.1
Completing of cut pieces	Hand makes	53	509	36.3
Control	Hand makes	77	351	52.7
TOTAL TIME:		443 s		

Technological operation plan for the sewing and finishing sweat is shown in Table 2.10.

Table 2.10 Technological operation plan for the production of sweat

Name of operation	Means of production	Pr quota/a piece (s)	Norma (piece)	Load (%)
Sewing middle seam internal and external hood (face and inside the hood)	SMo3	31	871	21.2
Sewing bar on external hood	OM	18	1500	12.3
Sewing cuff with the formation of openings for finger	OM	236	114	162.3
Turning cuff	RR	19	1421	13
Sewing belt on side	SMo3	5	5400	3.4

(Continued)

Operation	Machine			
Turning and bending the belt in half	RR	5	5400	3.4
Sewing shoulder to with shoulder reinforcement	SMo3	16	1688	11
Sewing external hood on neck part	SMo3	26	1038	17.8
Overloch seam front part for zipper and hole of internal hood	SMo3	17	1588	11.6
Sewing front left part with front part from zipper	OM	9	3000	6.2
Sewing zipper	OM	182	148	125
Sewing internal and external hood and neck part from the bar	SMo3	392	69	268.1
Turning hood	RR	19	1421	13
Closing neck part with bar	OM	75	360	51.4
Sewing sleeves	SMo3	46	587	31.5

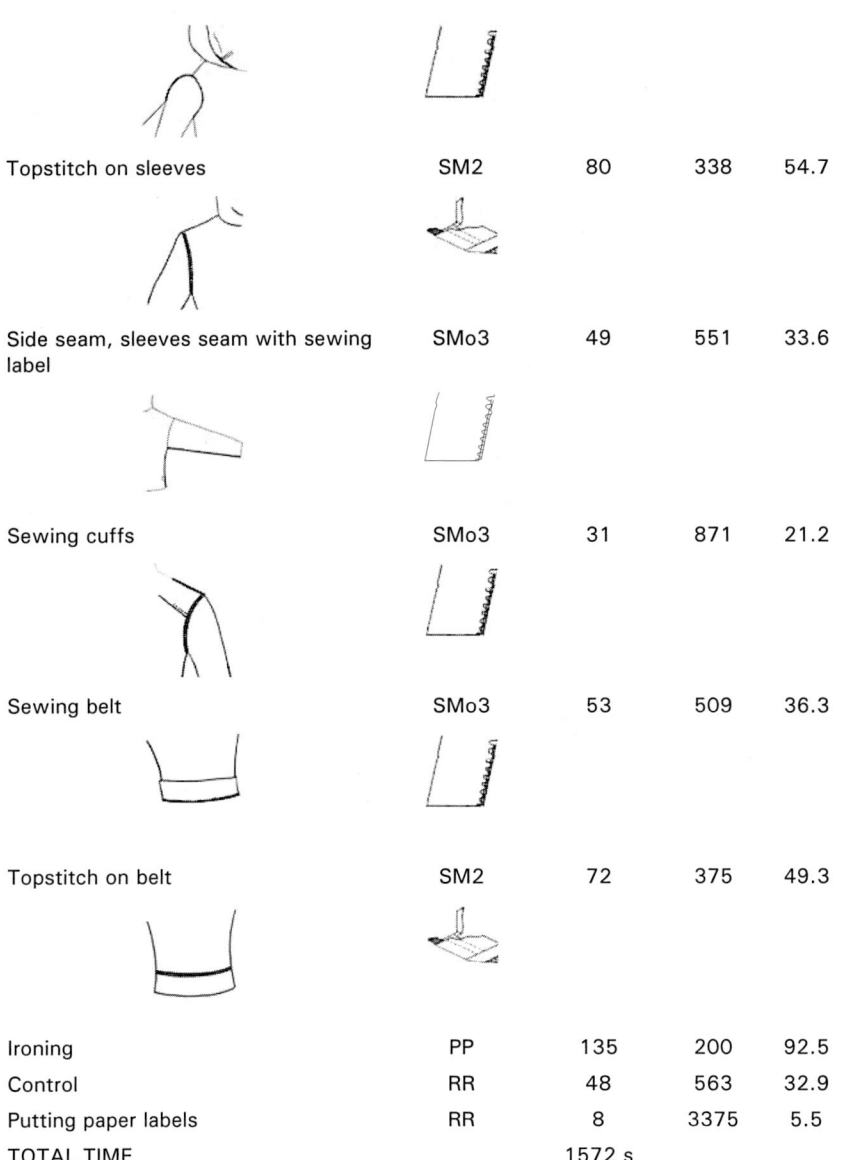

Topstitch on sleeves	SM2	80	338	54.7
Side seam, sleeves seam with sewing label	SMo3	49	551	33.6
Sewing cuffs	SMo3	31	871	21.2
Sewing belt	SMo3	53	509	36.3
Topstitch on belt	SM2	72	375	49.3
Ironing	PP	135	200	92.5
Control	RR	48	563	32.9
Putting paper labels	RR	8	3375	5.5
TOTAL TIME		1572 s		

References

1. Colovic G and Petrovic V (2001). The analysis of working time losses in a technological process of the production of men's T-shirts. *First International Ergonomics conference, Energonomy 2001*, Zagreb, pp. 143–153.

2. Colovic G, Paunovic D and Djordjevic J (2005). Using the technically reticular planning technical preparation of production apparels. *Fifth International Scientific*

Conference of Production Engineering, RIM 2005. Development and modernization of production, University of Bihac, pp. 759–764.

3. Colovic G, Paunovic D and Savanovic G (2009). Analysis of Classical and Modern Production Line for Production of Male Denim Jacket. *International Scientific Conference UNITECH 09*, Gabrovo, p. sp95.

4. Paunovic D and Colovic G (2004). *Prirucnik za konstruktore odece*, VTTS, Beograd.

5. Colovic G, Paunovic D and Djordjevic J (2005). 'The relations of anthropology characteristics and construction parameters for example children trousers', *5th International Scientific Conference of Production Engineering, RIM 2005, Development and modernization of production,* University of Bihac, pp. 765–770.

6. Paunovic D, Colovic G, Stojanovic O and Hotomski I (2007). Optimizacija proizvodnje odevnih predmeta po meri. *Seventh International Scientific Conference of Production Engineering, RIM 2007,* University of Bihac, pp. 203–208.

3

Determining time of technological operations in clothing production

Abstract: Time management of production is a comparative advantage for clothing manufacturers in global markets today, because it is reflected in the quality, cost and productivity of production. Therefore, it is necessary to make a thorough research of the structure of time of work and methods for determining the time of technological operations.

Keywords: time, methods, analysis of losses.

3.1 Methods for determining the time of technological operations in the production of clothing

All modern systems of production organization are based on the accurate and real time standards, stabilizing workplaces and methods of work and the rationalization of movement, i.e. following the model: Time – Accuracy – Quality.

The work study includes a variety of scientific methods of analyzing the work profile and production time, and allows finding the most economical work profiles and the time required by the average skilled worker with specific expertise to carry out a working operation, with normal effort.

Ralph M. Barnes (1947) used the name of the Time and Motion Study to emphasize that it is necessary first to define the appropriate method and then to determine the time. The time and motion study is defined as follows: "The time and motion study analysis methods, materials, tools and equipment used or to be used in the performance of a work – the analysis that is conducted with the intention to

- find the most economical way of performing this work;
- standardize the method, materials, tools and equipment;
- determine precisely the time required for the qualified and appropriately trained worker, who works at normal intensity, to do the task;
- assist in training workers for the new method."

Clothing production deals with norm working. Norm is the time the average skilful worker of appropriate expertise needs to perform a specific technological operation with normal effort and fatigue, at normal environmental action and under normal conditions of work. This definition suggests the following three conditions:

(1)   Qualified and well-trained worker;
(2)   Working at a normal pace with which a trained worker performs the task under normal conditions with a normal level of consumption of his own energy. Normal human consumption of energy is the one at which the employee can withstand the pace of work – neither too fast nor too slow. Carrying out of time norm that most of the skilled workers can perform is 100%.
(3)   Performing a specified task which has a certain prescribed method of work, materials specification, specification of tools, accessories and the equipment used, position of intermediate storage before and after performing the task, displaying additional requirements related to safety, quality, workplace regulations and activities to ensure the carrying out of the task.

Working norms vary according to the method of determining, how organization works and the method of expressing.

According to the method of determining, they can be

- technical – which are determined by the predefined standard elements of work,
- statistical – which are determined on the basis of statistical data, obtained over a longer period of time, by keeping a record of the work carried out,
- experiential – which are determined on the basis of experience and comparisons with norms for the same or similar jobs.

According to the method of work organization, they can be

- individual – these are characterized by the fact that the work is done independently by one person;
- group – these are characterized by the fact that the work is performed by two or more workers in the joint, same or similar work, which is not possible or is not rational to be separated per each individual;
- organizational units – these are characterized by the fact that the work is performed by all employees and units, i.e. when the job is not possible or is not rational to be separated per groups of workers or individuals.

According to the method of expressing, they can be

- Quantitative norms – number of pieces that should be made in a period of time, and the ratio of the daily working time (T_d) and time norm (e.g., pieces/h).
- Time norms – express the number of seconds, minutes, hours or days required for making one piece, operation or garment.

According to the number of workers, a single norm can be used for each position (e.g. for each operation in a sewing room) or a group norm for a group of workers in a particular phase of work where changes of models and even items frequently occur (e.g. in a cutting room, on final ironing – one employee often works on several presses for ironing). The joint effect is achieved by the participation of workers, the number of hours and the work category. Then a joint effect is shared among the individuals.

A time norm is most often used in garment industry. It helps in

(1) determining the cost of production, and thus determining the selling price of the product.
(2) developing the plan for machine engagement, the implementing manufacturing operations and determining workers necessary for the realization of production, delivery of materials and inventory management.
(3) balancing the assembly line and assembly line speed.
(4) specifying the qualifications and assessing the desired abilities and skills of workers.
(5) making payment of incentive pay.
(6) calculating the cost reduction according to given suggestions based on most economical method.
(7) calculation of feasibility procurement of new equipment in accordance with the height of production costs.
(8) monitoring performance in relation to the budget–wage and measuring results of leadership and management.

In case that some of the conditions, under which the norm was given, change (the technological process, stabilization of the workplace, machinery, materials, environment, etc.), a norm can be changed, but not the worker because he is not a condition of work. During the standardization, problems may occur because the norms are often set unrealistic because of

- Mis-prescribed norms.
- Miscalculated norms.
- Too high or too low on certain norms.

- Same norms, but new method of work.
- Roughly set norm, and so on.

Standardization of work is a sensitive area of management activity, so mistakes are very common to occur. The risk of error is twofold: if the standards are set too low and if they are set high. In both cases the work efficiency is being lost. In the first case, the worker will easily reach the target norm, perhaps after 3–4 hours of work, which means that he will achieve the full effect with half of his real possibilities. This damages the PBS, because the available human resources are not fully exploited. In another case, the worker will not be able to reach 100% of norm, he will be paid less, and if other positions in the same working process are properly normalized he will represent the so-called "bottleneck" of the process. Therefore, it is important to determine the required effort the worker can really put into 8 hours of work during the day. Different jobs require different efforts, and how much effort an individual will make, depends on the level of his work motivation. Thus, by measuring the effort (the highest possible performance for a specific work task and its variations), the level of work motivation of individuals can be measured as well.

Standardization of work and determining of standards are carried out through seven steps:

Step 1 – The choice of work.
Step 2 – Inform employees that the performance of their work is the subject of time study.
Step 3 – Divide the work into smaller elements.
Step 4 – Determine the size of the sample.
Sample size is determined by formula:

$$n = \left[(\frac{z}{a})(\frac{s}{x}) \right]^2 \tag{3.1}$$

Where: z = number of standard deviations with the corresponding error risk,
s = standard deviation of the sample,
a = degree of accuracy,
$\bar{x}$ = average value of the sample.
Step 5 – Measure the time of performing each element of work.
Step 6 – Calculate the usual time.
Step 7 – Assess the standard time.

Technological processes of clothing production belong to the piece manufacturing with the "chain systems" of a production flow of materials, so the time basis for determining the norms of production capacities of

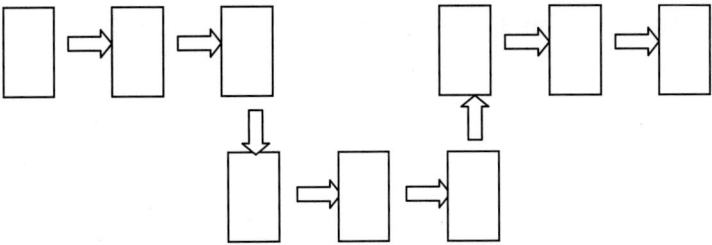

3.1 Schematic diagram of the chain system.

workplaces and successful organization of production are very real. The "chain system" of manufacturing involves the performance of working operations one after another in a specific order from one workplace to another, at a certain time (rhythm) and with the shortest spatial area. This system (Figure 3.1) doesn't have any intermediate storages and it is suitable for products with a smaller number of work operations where changes of models are not frequent. Advantages of "chain systems" are small work space, fast flow of materials, and low production of unfinished and less engaging financial resources. Disadvantages of "chain systems" are sensitivity to frequent changes of model, the lack of workers causes a delay (spare worker is needed), individual ability of workers cannot be seen and maximum productivity is not achieved.

The main characteristic of the chain system is moving of the object of work following the already defined rhythm ("tact in" German) regardless of the transport of the object of work. According to the method of transporting the subject, we differ human work linked by rhythm and human work caused by rhythm.

A rhythm of a group means an average time of keeping the object of work on a working place and means a maximum allowed time load of a workplace in human work linked by rhythm. Shift related to the human work linked by rhythm is related to the speed of moving of transport. The worker must be in the specified time (rhythm) to perform the operation, because otherwise there is a "bottleneck" for production or waiting to work. In a system with continuous motion a conveyor speed must be adjusted to the time of making the object of work. Time of the pass of work object is calculated from:

$$T_{pr} = \frac{t_1}{ORO} \cdot 100 \qquad [3.2]$$

Where: t_1 – the production time,
ORO – working load operation.

Thus rhythm of a group is calculated from the formula:

$$t_g = \frac{T_{pr}}{R} \text{[minutes]} \qquad [3.3]$$

Where: R – the number of workers.

In qualifying the human work caused by rhythm with inserting bundles (packages), a rhythm of a group of workers can be determined on the basis of the average production time and number of workplaces, because the differences are undermined by helping slower workers, i.e. it is calculated from the formula:

$$G = \frac{t_1}{R} \qquad [3.4]$$

To achieve the continuity of production, it is necessary to control the rhythm of a group (e.g. every 2 hours). The rhythm of production causes a range of problems. A worker, given the demands of high training, does the same operations for years and his work becomes monotonous. As a result of this manufacturing strategy, various occupational diseases may occur, as well as injuries, disability, and increased fluctuation of employees. To overcome such phenomena, the establishment of group and team work is required. Thanks to the introduction of rotating workplaces within the group, additional effects are achieved. A rotating workplace allows introduction of innovations in production and reduction of work load, and it gives the quality of production.

With small series, work operations are performed uniformly in order to increase productivity and quality of clothing. So the overall costs of production are therefore reduced and they are normed easier, i.e.

$$y = T_b + \frac{T_0 - T_b}{\dfrac{x}{2n}} \text{[minutes]} \qquad [3.5]$$

Where: T_0 [minutes] – initial time of duration of operation,
T_b [minutes] – minimum time of duration of operation,
n – number of repeated operations when the time is reduced to half of the previous time,
x – number of repetitions of the same operation.
Figure 3.2 shows the continuity of production.

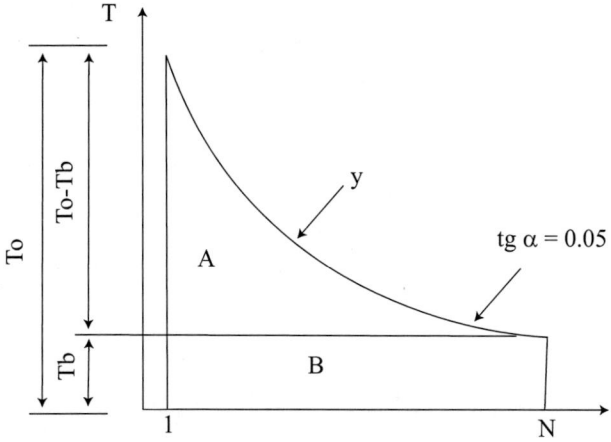

3.2 Continuity of production ("well trained" production).

3.2   Determining the production time

Normal production time and labour standards are an important tool for solving many problems in the work of a PBS. The quality of planning, the relationship among employees, offers, contracts etc., depends on the reality of normal production time. Establishing normal production time allows

- Planning and scheduling of production.
- Estimation of earnings.
- Calculation of the earnings performance.
- Deciding about cooperation.
- Calculation of buying the machinery and equipment.
- Calculation of delivery deadlines.
- Calculation of costs of production.

To obtain the normal production time it is necessary to analyze the constituent elements of operations, such as grips, movements and micro-movements (Figure 3.3).

The term operation implies the processing of cases on one machine, or at one workplace. An operation consists of all the works from the moment of its starting point to the moment of its storing on the planned location. Operation is then divided into grips.

A grip is a direct effect of tools on the subject of work during the operation. Grips can be of the following types:

(a) Pre-final grips – these include activities related to the preparation of workplace and workers to perform the operation, activities which

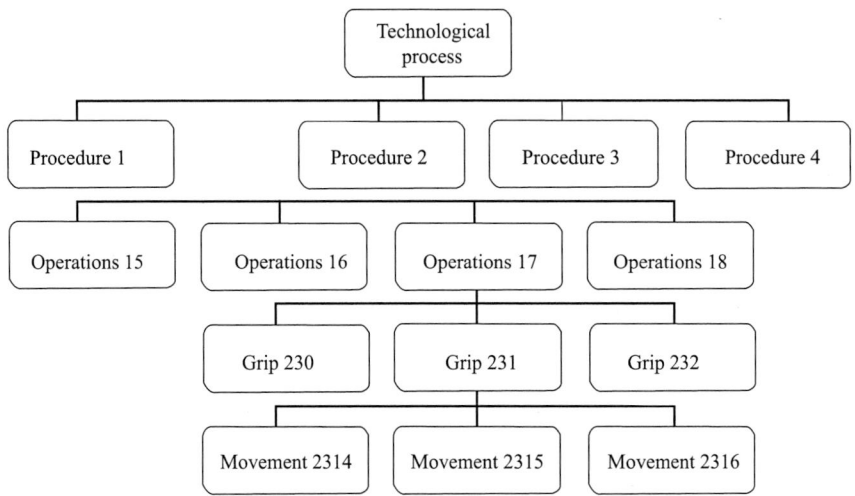

3.3 Elements of technological process.

deal with cleaning up the workplace after the operation was finished; receiving and studying the technical and technological documentation; receiving and review of material; submitted documents; receiving and replacement of standard tools and supplies. These grips are performed once during the performance of operation. They are most commonly given in the table depending on the type of machine and the type of production.

(b) Additional grips – these include all ancillary activities in the workplace and on the machine itself which allow the performance of grips (placing and removal of work object on the machine, positioning of work objects, setting certain parameters of machine, switching the machine on and off, etc.). They are most commonly given in the table depending on the type of machine and the type of production.

(c) The processing grips – they are the direct processing of work object, and they can be elementary grip processing, complex grip processing and group grip processing.

In elementary grip processing a certain surface on the work object of processing is formed by one tool and its moving (e.g. drilling holes, cutting holes with a knife, putting rivets on garment). Complex grip processing is a part of the processing operation where one tool is used for the final formation of a complex area according to technical–technological requirements. Group grip processing is a completely or partially simultaneous process of forming a number of surfaces with a number of appropriate tools according to technical–technological requirements, where certain elements of the process may be the same or different (embroidery machine).

Movement is the element of grip and includes an activity of workers with tools or accessories. Movement is, for example, scrolling tools, cutting threads, scissors staying away, changing direction of rotation of the work object, raising pedals, etc. Movements can be divided into micro-movements that can not be split into smaller parts.

The structure of a component time required for performing an operation (Taborsak, 1971) consists of production time, additional time and a pre-final time.

(1)   *Production time* (t_{iz}) includes the time required from the first to the last movements of the work of employees. It consists of a technological time (t_t) and help-time (t_p). Technological time is the time required for a specific operation, and may be machine-technological time (on sewing machines), machine-hand technological time (on ordinary sewing machines and special sewing machines) and hand-technological time (for hand makes). Distinguishing these times, by Martinovic (2002), is necessary because of the following:

- Replacing of machine and hand works in one operation (allows employees to work on two machines).
- Large taking part of machine work in performing an operation (30%) allows employees to work on many machines.
- High-speed working machines due to which grips and movements related to the work of machines cannot be corrected.
- Abilities to calculate a machine-technological time and a machine-hand technological time mathematically and not to perform their measurements.

Machine-hand technological time in the process of sewing a meter of seam is determined by the formula:

$$t_1 = \frac{\sum_{i=1}^{n} t_{ui} + \sum_{i=1}^{n} t_{ni} + \sum_{i=1}^{n} t_{ki}}{P\left(\sum_{i=1}^{n} \int_{t_{pz}}^{t_{ppz}} (t)dt\right)} \cdot 1000 \qquad [3.6]$$

Where: n – the number of machine and machine-hand technological grips,

P [mm] – transport of object of work or length of stitch,

t_{ui} [s] – time acceleration of the main shaft of sewing machines,

t_{ni} [s] – working time of sewing machines,

t_{ki} [s] – time of slowing down the main shaft of sewing machines.

Help-time is the time required for grips that enable and assist in performing the process (e.g. switching machines on, control during working, replacement of sewing needles, thread changes, lubrication and cleaning of machines, etc.).

(2) *Additional time* (t_d) – it is the time of the loss of working time which does not depend on the worker himself, and is expressed by three coefficients: fatigue coefficient (K_n), the coefficient of environmental influence (K_a) and the supplementary coefficient (K_d).

The fatigue coefficient is required because of the reduction of performance due to physical and mental fatigue of workers during working time caused by monotony, hard work, body position, etc.

With physical fatigue, as a consequence of dynamic muscle strain, exhaustion and fatigue may occur (if the work is limited to some constant movements and muscle groups), and as a result of static muscle strain (if the limbs are immobile and if there is a permanent contraction of muscle groups). Muscle fatigue is a lack of oxygen, and occurs whenever a bigger impact within a unit of time is required than it corresponds to the amount of oxygen that is available.

Normal positions of the body while working are standing, while all others require higher power consumption. The most appropriate one is the sitting position, because then the man spends only 5% more energy than when lying. Standing requires more static muscle strain, that's why the power consumption is 10–15% higher.

The coefficient of environmental influence is the activity caused by inadequate temperature, relative humidity, steam, noise, etc.

Supplementary coefficient is required because of the prescribed resting of 30 minutes, which is the time required for the personal needs of employees and organizational losses, e.g. waiting for the material. For determining of supplementary coefficient, about 15% are taken, including (Adamovic and Alihodzic, 2002):

- 5% for personal needs,
- 4% for interruptions caused by personal fatigue,
- 2% for a very unfavourable position of the body,
- 2% for insufficient light,
- 2% for precise work.

(3) *Pre-final time* (t_{pz}) is the time required for preparation of workplace and its cleaning after finishing the work, and it's divided into:

(a) preliminary time (t_{pr}) – the time required for preparation of machines and tools, i.e. the workplace before and during the work, preparation of materials, and receiving and reviewing of documents,

(b) final time (t_z) – the time required for cleaning of materials and cleaning of machines and tools.

It is still a characteristic in garment industry that 80% of time is spent on handling the case work, assembling of parts of clothes, positioning and alignment, and only 20% of time is used for performing of machine and machine-hand grips.

In order to record the time on a proper way it is necessary to be prepared for recording, which includes the following activities:

(1) Analyze selected operations through grips, movements and micro-movements, in order to specify the parts of the operation whose duration should be recorded.

(2) Determine which parts of the operation are repeated and which are specific; separate machine work from hand work; define the start and the finish.

(3) Make selection of methods and recording tools.

(4) Recording of conditions under which the operation is carried out, because working conditions are subject to change.

(5) Selection and preparation of workers – a number of workers of different abilities, or a worker of average ability.

(6) The choice of number of recording on the basis of statistics.

Production time is determined using several conventional methods:

(a) *Continuous time method* uses a stopwatch timer with hour mechanism in which a minute is divided into hours, and hour into 10000 parts. There are two large watch hands, with only one of them moving smoothly during the recording, and the other one which can be stopped (e.g. when it is necessary to wind the lower thread on spool). The recorder takes only a total time and calculates the average production time.

Advantages of this method are as follows: it measures the total time of entire operation; working time is not wasted; there's no need for another watch for control; training of workers doesn't last long.

Disadvantages of this method are as follows: it's difficult to track a worker and record if grips in operation are bypassed; it can not determine the error of chronometer; it's difficult to record grips shorter than five seconds.

Stopwatch time study was developed by Frederick W. Taylor in 1880 and it was the first technique to be used to set the engineering time standard.

(b) *Snapback time method* uses a stopwatch timer, where a large dial is divided into one hundred parts, each of them being a one hundredth part of a minute

3.4 Stopwatch timer and electronic stopwatch.

worth. When a big watch hand makes full circle, then a small watch hand on a small dial moves for one degree and shows the value of one minute.

Advantages of the method are as follows: it's possible to identify all the errors during work; there's no subsequent calculation of individual times; the employee may skip the grips because they are noted separately; it's possible to record short grips.

Disadvantages of this method are as follows: error must not be bigger than 1.5% so the recorder must practice more; quick reflexes are required as well as large concentration of a recorder; there are losses due to the return of the watch hand to the starting position.

In Figure 3.4, the stopwatch timers are shown.

(c) *Three-Watch Time Study* is a better technique than both the continuous and the snapback ones. Three continuous stopwatches are used on one board for each stage. When the worker finishes the grip, the recorder pulls a common switch and presses it down. One watch is stopped so a reading can be made, the second watch is restarted and the third watch is reset to zero waiting for the time of the next grip.

(d) *Method of recording with film and video cameras* is used for technological grips and movements of very short duration, which requires a large measurement accuracy.

(e) *Measurement method using the computer* – there are several modern methods, for example the method of measuring process parameters (MMPP) is based on measuring the time of the main shaft speed and the calculation of real and average sewing speed, acceleration, number of stitches in the seam, etc.

Basic procedure of measurement consists of three phases:

(1) Analysis – this is the phase in which the job is divided into smaller sections, i.e. procedures, elements;
(2) Measurement – this phase refers to measuring the time necessary for a qualified employee to perform each element within a particular job under certain conditions;
(3) Synthesis – in addition to the basic time required for a particular job, displaying the total time required for a particular job is done in this phase. The basic time, defined in the measurement phase, is added to the time for resting, for possible errors, etc.

A snapback time method with stopwatch timer is most widely used in garment industry. The calculated time is multiplied with the coefficient of efforts which gives the normal production time (t_n):

$$t_n = t_0 \cdot K_{pz} = t_0 \cdot \frac{p_z}{100} \qquad [3.7]$$

or

$$t_n = \frac{1}{n} \sum_{i=1}^{n} t_0 \cdot P_z \qquad [3.8]$$

Where: t_0 – the average value,
       K_{pz} – the coefficient of efforts (rating factor),
       p_z – the constant.

Assessing the rating factor is determined by the scale which ranges from 70% to 130%, with five percent leaps (e.g. 70%, 75%, 80% ...).

However, a worker must increase the strain while carrying out work tasks. Therefore, it is necessary to add the time which will make up these elements of non-working. Additional time (t_d) is a time of loss of working time which does not depend on the worker (expressed by three coefficients: the fatigue coefficient, the coefficient of environmental influence and the supplementary coefficient). The real time (t_s) is obtained through the formula:

$$t_s = t_n \cdot (1 + K_n \cdot K_a) \qquad [3.9]$$

The real time may be technological time and help time:

$$t_{ts} = t_n \cdot (1 + K_n \cdot K_a) \qquad [3.10]$$

$$t_{ps} = t_n \cdot (1 + K_n \cdot K_a) \qquad [3.11]$$

And since the production time is, by definition, equal to the sum of technological time and help time:

$$t_i = t_t + t_p = t_{ts} + t_{ps}$$ [3.12]

The real time is equal to the production time on the basis of which we get the time norm:

$$T = t_i \cdot (1 + K_d)$$ [3.13]

As an example, the times recorded in the factory for production of sports clothing are given. In the Table 3.1 there are the results of the observed time of making the shoulder seam on a special sewing machine (SM-o5) with continuous time method.

Table 3.1 Results of recording production time of a shoulder seam on men's polo shirt

Time

dmh*	h	Number of notes	Σ Notes	Coefficient of efforts (%)
91	0.0091	IIII	4	130
94	0.0094	IIIIIIII	9	125
97	0.0097	IIIIIIIIIIIIIII	7 + 9	120;115
100	0.0100	IIIIIIIIIIIIIIIIIIIII	22	110
102	0.0102	IIIIIIIIIIIIIIIIIIII	10 + 10	105;100
105	0.0105	IIIIIIIIIIIIIIIIIIIIIII	23	90
108	0.0108	IIIIIIIIIIII	12	80
111	0.0111	III	3	75

dmh units = HOUR/10 000

According to Table 3.1, normal production time (basic time) of shoulder seam on a special machine, according to the formula [3.7], is $t_n = 0.01$. If the fatigue coefficient of worker is $K_n = 0.11$ and the coefficient of environmental influence is $K_a = 1.3$, then the real time is calculated according to the formula [3.8] $t_s = 0.01 \ (1 + 0.11 \cdot 1.3) = 0.01143$. The real time equals the time of production ($t_s = t_i$). If the supplementary coefficient (extra time) is $K_d = 0.15$, then the norm is calculated according to the formula [3.13]: $t_1 = 0.01143 \ (1+0.15) = 0.01322$ [hour/operation].

In order to calculate the time norm other methods can be used, such as (Adamovic and Alihodzic, 2002):

(1)   Michelin methods – they are used to calculate the time norms according to which it is necessary to verify the number and quality of recording (observing) operations. The results sorted in the order of recording are divided into two equal groups, Ma and Mb, and the arithmetic mean of Ms is calculated.

$$M_s = \frac{Ma + Mb}{2} \qquad [3.14]$$

Then the deviation of allowable 5% is determined and, according to this method, the normal time is obtained as the arithmetic mean of the minimum (Tiz_{min}) and middle (Tiz_{sr}) production time. This time is called "the standard production time" (Tiz_{tip}) and it is calculated from the formula:

$$Tiz_{tip} = \frac{Tiz_{min} + Tiz_{sr}}{2} \qquad [3.15]$$

Pre-final time is calculated from the formula:

$$T_{pz} = 0.2 \cdot Tiz_{tip} \qquad [3.16]$$

Additional coefficient:

$$T_d = 0.15 \cdot Tiz_{tip} \qquad [3.17]$$

Therefore, a time norm is obtained from the formula:

$$T_{pn} = T_{pz} \cdot T_d \cdot Tiz_{tip} \qquad [3.18]$$

(2)   The method of average value – it is similar to the previous method because it is based on the arithmetic mean of obtained results. The mean value, the minimum value and the standard production time are the same in this method.

(3)   The method of maximum frequency – it takes the value of maximum frequency as a normal production time and other data are obtained as in the Michelin method. For example, for one operation the following times are recorded in this order: 6,7,8,8,9,9,9,10,11,12,13. On the basis of this a chart for the frequency of time is formed (Figure 3.5), which clearly shows that 9 time units is the most frequent time and it usually appears three times during the measurement. Therefore, the frequency is f = 3,

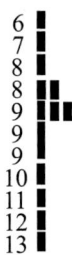

6
7
8
8
9
9
9
10
11
12
13

3.5 Frequency of time.

and T = 9 (time units, e.g. seconds).

Michelin method is most suitable for the owners, and the method of maximum frequency for the workers.

Table 3.2 Results of observed time of sewing placket on the front of the men's polo shirt on a sewing machine

Time,s	Number of notes	Σ Notes	Coefficient of efforts (%)
33	II	2	130
34	III	3	125
35	II	2	120;115
36	IIIIII	6	110
38	III	3	105;100
39	II	2	96
40	II	2	80
	Σ	20	

As an example, Table 3.2 shows the results of recording time of the first operation of producing men's polo shirts on a sewing machine.

On the basis of the results of recording, four methods for calculating working norms were used:

(1) The method for calculating working norm with the coefficient of worker's efforts:

Normal production time is t_n = 38.89s. If the fatigue coefficient of workers is K_n = 0.11 and the coefficient of environmental influence is K_a = 1.3, according to the formula the real time is t_s = 38.89 (1 + 0.11 · 1.3) = 45.59 s. The real time equals the production time ($t_s = t_i$). If the additional coefficient (extra time) is K_d = 0.15, then the production time is t_1 = 45.59 (1+0.15) = 52.46 s = 0.0145 [hour/operation].

(2) The Michelin method for calculating time norm:

The obtained results are divided into 2 groups and their arithmetic means are calculated, as well as the arithmetic mean of the set Ms:

$$Ma = 35.8 \text{ s}; \quad Mb = 36.8 \text{ s}$$

$$Ms = \frac{35.8 + 36.8}{2} = 36.3_s$$

In order to determine the quality of results, the arithmetic mean of set is increased by five percent (or $0.05 \cdot 36.3 = 1.815$) and the deviations are determined:

$$| \text{Ms Ma} | = | 36.3 - 35.8 | = 0.7 < 1.815$$
$$| \text{Ms Mb} | = | 36.3 - 36.8 | = -0.5 < 1.815$$

Recorded times are divided into three approximately equal groups and the biggest, i.e. the last value from the first group is taken as the minimum time $\text{Tiz}_{min} = 35$ s. Since $T_{sr} = \text{Ms}$, normal (standard) production time is

$$\text{Tiz}_{up} = \frac{35 + 363}{2} = 36.65\,\text{s} = 0.010\,[\text{hour/ operation}].$$

Therefore, pre-final time and additional coefficient are

$$T_{pz} = 0.2 \cdot 36.65 = 7.99\,\text{s} = 0.002\ [\text{hour/operation}]$$
$$T_{d} = 0.15 \cdot 36.65 = 5.99\,\text{s} = 0.0015\ [\text{hour/operation}]$$

According to the Michelin method, the norm for the first operation of sewing men's polo shirt is

$$T_{pn} = 0.002 \cdot 0.0015 \cdot 0.010 = 0.0135\ [\text{hour/operation}]$$

(3) The method of average value:

On the basis of the obtained results the arithmetic mean, which is the minimum value at the same time, is calculated: $\text{Tt}_{min} = 36.3$ s; $\text{Tiz}_{sr} = \text{Tt}_{min} = 36.3$ s. Normal (standard) production time is $\text{Tiz}_{tip} = 36.3$ s $= 0.01016$ [hour/operation]. As in the Michelin method:

$$T_{pz} = 0.2 \cdot 36.3\,\text{s} = 7.26\,\text{s} = 0.002\ [\text{hour/operation}].$$

$$T_{d} = 0.15 \cdot 36.3\,\text{s} = 5.99\,\text{s} = 0.0015\ [\text{hour/operation}]$$

The norm, according to this method, is

$$\text{Tpn} = 0.002 \cdot 0.0015 \cdot 0.0101 = 0.0136\ [\text{hour/operation}]$$

(4) The method of maximum frequency:

The Table 3.2 shows that the value of the highest frequency is considered a normal production time (36 seconds are repeated six times):

$$\text{Tiz}_{tip} = 36\,\text{s} = 0.01\ [\text{hour/operation}]$$
$$T_{pz} = 0.2 \cdot 36\,\text{s} = 7.2\,\text{s} = 0.002\ [\text{hour/operation}]$$
$$T_{d} = 0.15 \cdot 36\,\text{s} = 5.9\,\text{s} = 0.0015\ [\text{hour/operation}]$$

According to the method of maximum frequency, the standard is $T_{pn} = 0.002 \cdot 0.0015 \cdot 0.01 = 0.0135\ [\text{hour/operation}]$.

The values of calculated time norms, according to four methods mentioned above, are shown in Table 3.3. The obtained results indicate the variations in production planning.

Table 3.3 Values of norms according to different methods

| Method | Time norm | | Working time (hour) | Quantitative norm - number of pieces |
	s	hour/operation		
1.	52.2	0.0145	7	483
2.	48.6	0.0135	7	518
3.	48.9	0.0136	7	514
4.	48.6	0.0135	7	518

3.3   MTM method

Increasing demands for productivity encourage scientific researches in the field of work study, in order to find procedures and methods that would ensure required performances without any additional efforts. The important place among the scientific methods for determining the optimal way of performing the work process belongs to the Methods Time Measurement (MTM), developed in the 1940s by Maynard, Schwab and Stegemerten, which was later called the MTM-1.

MTM system is the most famous one in the world and it is mostly used in garment industry. It is used for the rationalization of existing work procedures and for providing objective data to design new work processes. This system is a time system which divides every manual technological operation into basic movements which are given certain time values depending on the nature of movements and the conditions in which they are carried out. MTM system consists of tables of time for movements: 9 basic movements of fingers, hands and arms, 2 eye movements, 10 movements of the body, legs and feet with about 400 standard time norms of basic movements. Basic movements of MTM system, used in garment industry, are

(1) Reach (R) – stretching out fingers, hands or arms to a certain position or an object of work. Time amounts of movements depend on the length of movement, dynamics of performance (type 1, 2, 3) and position of object of work (A, B, C, D, E).
(2) Grasp (G) – establishing control and contact with the object of work.
(3) Move (M) – changing the position of object of work.
(4) Release (RL) – termination of control over object of work.
(5) Position (P) – when the hand or fingers with one or more micro-movements put the object of work into a precisely defined position.

(6) Turn body (TB) – swivelling body 45° to 90° with the relocation of one or both legs.

(7) Walk (W) – moving body from one position to another.

Duration of movement is given in units of TMU (Time Measurement Unit) with 1 TMU = 0.00001 h = 0.036 s.

Standard time in the tables does not include extra time for physiological needs, resting and other relevant conditions (micro-climate, noise, body position, etc.).

The procedure of applying MTM (Maynard and Schwab, 1984) occurs in the following order:

(1) Selection of technological operations on the basis of analysis of technological operation plans for sewing and finishing (for example, analysis of Table 2.6).

(2) A detailed description of all operations on the basis of the obtained data (type of seam, seam length, type of stitches, density of stitches, etc.) and detailed determining of all relevant parameters for machine, tools and accessories, the object of work, working conditions and requirements in terms of quality.

(3) Designing of workplaces on the basis of technological, technical, ergonomic and economic principles.

(4) Rational division of technological operations onto grips.

(5) Determining of basic action movements.

(6) MTM analysis of basic movements (body, arms, legs, feet and eyes).

(7) MTM analysis of the performance capabilities of combined movements and all movements that can be performed simultaneously.

Combined movements are a set of basic movements, which are simultaneously performed by the same part of the body (left leg, right leg, left arm, right arm, torso and head). A set of basic arm movements (fingers, hand, forearm and upper arm) allows combining of performances such as reaching, moving, grasping, swivelling, pressing and disassembling. A set of basic movements of the head contains the possibility of turning the head and eyes in the horizontal (± 55°) or vertical direction (± 45°).

Simultaneous movements are performed simultaneously with different parts of the body (with both hands, arms and legs or both hands and a foot) and allow less fatigue of workers, greater speed and accuracy in work. According to their character, simultaneous movements can be identical, similar and different movements, and they are systematized into three classes:

Class I – basic movements that are easy to perform simultaneously.

Class II – basic movements that are performed simultaneously, but with the necessary training.

Class III – basic movements that can not be performed simultaneously, so they are analyzed separately.

(8) Defining work methods and determining the amount of time for the operations, grips and movements.

(9) Calculating the cumulative time for working elements, adding extra time (in case of working on a machine the possibility of simultaneous performance of certain work operations with machine work is considered) and expressing in TMU.

(10) Determining of normal production time.

(11) Checking up whole procedure so as to correct possible errors.

Example of MTM analysis for making six straight holes on sewing automatic machine on the front of men's shirt is given in Table 3.4.

Table 3.4 MTM analysis for making six straight holes on sewing automatic machine on the front of men's shirt

Left hand	Symbol	TMU*	Symbol	Right hand
	1. Taking the pieces of garment			
reaching the front part	R30B	14,2	R20B	reaching the front part
			G5/G2	taking the front part
taking the front part	G5	8,8	M15B	up the front part
reaching the bordures of front part	R15B			
taking the front part	G1A	2,2		
	2. Putting the front part on automat			
up the front part	mM10B	4,3	mM10B	up the front part
putting on the machine	M30A	12,7	M30A	putting on the machine
closing tapes	M45B	16,8		
up the part of machine	M10A	6,0		
down the front part	RL1	2,0	RL1	down the front part
balance of the front part(twice)	R10B	12,6	R10B	balance of the front part(twice)
taking the front part	G5	2,0	G1A	taking the front part
	3. Position of the front part on the sewing machine			
		6,8	M10B	taking to the stitch place
		5,8	M6C	putting on the stitch place
		16,2	P2SE	putting on the mark
		2,0	RL1	down the front part

		15,6	R40B	reaching the front part
		0,0	G5	taking the front part
down the front part	RL2	0,0		
taking the switch	R20A	7,8		
switching on machine	G5/AF	3,4		
4. Making 6 button holes with automatic movement				
machine work		739,0		
5. Walking to next machine				
rotating body at 90°	TB2	37,2		
rotating body at 45°	TB	18,6		
walking	WM1,5	26,1		

Where:

TMU — Time Measurement Unit $= 10^{-5}$ h $(3.6 \cdot 10^{-2}$ s),

R30B — reaching work object whose position can be changed (the length of movement of 30 cm),

R20B — reaching work object whose position can be changed (the length of movement of 20 cm),

G5/G2 — grasping work object by touch and re-grasp,

G5 — grasping work object by touch,

M15B — moving work object to the indefinite position by the length of movement of 15 cm,

R15B — reaching up object whose position can be changed (the length of movement of 15 cm),

G1A — grasping work object that is easy to grasp,

mM10B — moving work object,

M30A — moving work object to delimiter,

M45B — moving work object to the indefinite position,

M10A — moving work object to the delimiter (the length of movement of 10 cm),

RL1 — dropping work object,

R10B — reaching work object whose position can be changed (the length of movement of 10 cm),

M6C — moving work object to a specific position,

RL2 — placing work object,

G5/AF — grasping work object by contact touch and pressing by force (by finger),

TB2 — rotation of body of worker for 45° to 90° with the relocation of both legs for balance,

TB — rotation of body of worker for 45° to 90 ° with the relocation of one leg,

WM1.5 — step-walking 1.5 m.

Garment industry also uses the German MTM Association Table (Table 3.5) on the basis of which the sewing speed is determined (v_s) by reading values for stitches sewing speed (v_b) and the density of stitches (g_b).

Table 3.5 MTM

values stitches sewing speed v_b (stitches min^{-1})							sewing speed v_s	sewing time t_t/TMU seam length l_s (cm)					
density of stitches g_u (stitches cm^{-1})													
2	2,5	3	3,5	4	4,5	5	2	5	10	20	30	40	50
40	50	60	70	80	90	100	2	175	425	842	-	-	-	-
80	100	120	140	160	180	200	4	92	217	425	842	-	-	-
120	150	180	210	240	270	300	6	64	147	286	564	842	-	-
160	200	240	280	320	360	400	8	50	113	217	425	633	842	-
200	250	300	350	400	450	500	10	42	92	175	342	509	675	842
240	300	360	420	480	540	600	12	36	78	147	286	425	564	703
280	350	420	490	560	630	700	14	32	68	128	247	366	485	604
320	400	480	560	640	720	800	16	29	61	113	217	321	425	530
360	450	540	630	720	810	900	18	27	55	101	194	286	379	472
400	500	600	700	800	900	1000	20	25	50	92	175	258	342	425
440	550	660	770	880	990	1100	22	24	46	84	160	236	312	388
480	600	720	840	960	1080	1200	24	22	43	78	147	217	286	356
520	650	780	910	1040	1170	1300	26	21	41	73	137	201	265	329
560	700	840	980	1120	1260	1400	28	20	38	68	128	187	247	306
600	750	900	1050	1200	1350	1500	30	20	36	64	120	175	231	287
640	800	960	1120	1280	1440	1600	32	19	35	61	113	165	217	269
680	850	1020	1190	1360	1530	1700	34	18	33	58	107	156	205	254
720	900	1080	1260	1440	1620	1800	36	17	32	55	101	147	194	240
760	950	1140	1330	1520	1710	1900	38	17	30	52	96	140	184	228
800	1000	1200	1400	1600	1800	2000	40	17	29	50	92	134	175	217
840	1050	1260	1470	1680	1890	2100	42	17	28	48	88	128	167	207
880	1100	1320	1540	1760	1980	2200	44	17	27	46	84	122	160	198
920	1150	1380	1610	1840	2070	2300	46	17	27	45	81	117	153	190
960	1200	1440	1680	1920	2160	2400	48	17	26	43	78	113	147	182
1000	1250	1500	1750	2000	2250	2500	50	17	25	42	75	108	142	175
1100	1375	1650	1925	2200	2475	2750	55	17	24	39	69	99	130	160
1200	1500	1800	2100	2400	2700	3000	60	17	22	36	64	92	120	148
1300	1625	1950	2275	2600	2925	3250	65	17	21	34	60	85	111	137
1400	1750	2100	2450	2800	3150	3500	70	17	20	32	56	80	104	128
1500	1875	2250	2625	3000	3375	3750	75	17	20	31	53	75	97	120
1600	2000	2400	2800	3200	3600	4000	80	17	19	29	50	71	92	113

1700	2125	2550	2975	3400	3825	4250	85	17	18	28	48	67	87	107
1800	2250	2700	3150	3600	4050	4500	90	17	18	27	46	64	83	101
1900	2375	2850	3325	3800	4275	4750	95	17	17	26	44	61	79	96
2000	2500	3000	3500	4000	4500	5000	100	17	17	25	42	59	75	92
2200	2750	3300	3850	4400	4950	5500	110	17	17	24	39	54	69	85
2400	3000	3600	4200	4800	5400	6000	120	17	17	22	36	50	64	78
2600	3250	3900	4550	5200	5850	6500	130	17	17	21	34	47	60	73
2800	3500	4200	4900	5600	6300	7000	140	17	17	20	32	44	56	68
3000	3750	4500	5250	6000	6750	7500	150	17	17	20	31	42	52	61

Sewing speed can be calculated by the formula:

$$V_s = \frac{0.1 \times v_h}{g_h}$$ [3.19]

Sewing time in TMU is calculated from the formula:

$$T_t = 4.63 \cdot v_s$$ [3.20]

In terms of large series and mass production, as well as in cases of highly repetitive operations in administration, commerce and other service industries the usage of MTM is justified. Systematic analysis of the work process with the application of MTM in garment industry makes it possible to

- Specify the duration of technological operations.
- Determine the rhythm of the group.
- Establish the optimal method of performing work.
- Rationalize the existing procedures of work.
- Plan the technological process of clothing optimally.
- Determine the real work norm.
- Organize work on several machines.
- Determine the real costs of apparel production.

For small series production and individual activities which require high speed of analysis, MTM-2 and MTM-3 were developed, where a single logical set of basic movements create standard grips.

MTM-2 uses the reduced structure of basic movement and its variant cases. Compared to MTM, MTM-2 provides the results of deviations up to 15%, but the usage is simple, fast and cheap (it can be used in garment assembling or in the warehouse).

The variant of MTM-3 is a further simplification related to the MTM-2. Differentiation of movements is reduced to:

- Object handling and management by hand or fingers in order to put it in a new position.

- Transport (placing of object by hand into a new position).
- Step made by leg.
- Bending with the correction.

With the development of computer support in the field of work measurement, MTM Association has developed a 4M system (Micro-Matic Methods and Measurements). The 4M identified two sets of movements, such as:

- GET – reach with inclusion in appropriate combinations and
- PLACE – set on the place with possible variations referring to the fixing of object of work.

Software package for the application of 4M makes getting of reports possible, i.e.:

(1) Analysis of the operation which includes, for each of the two basic elements, the corresponding variation of time for each hand, total allowed time and normal production time.
(2) Instructions for worker how to do operations.
(3) Time for left and right hand, total time and performance.
(4) Analysis of operations with sets of code movements and their variations.

3.4   Method of relationship between the speed of forming stitches and time

Within the production of clothing, in the structure of technological operations of sewing, only 20–30% of machine and machine-hand grips are used, and they determine the technological time of sewing on the machine. The biggest problem is to determine the machine-hand times, because the movements of workers are affected by different factors, such as: technical characteristics of the sewing machine, type of seam, type of stitches, seam path curvature, psycho-physiological and cybernetic characteristics of workers, random grips when a needle is broken, material processing, short delays and so on. When performing machine grips, time is specified by construction of a sewing machine and it does not depend on the worker, so the fatigue coefficient, the coefficient of environmental influence and the supplementary coefficient are not added.

The researches in garment industry, which have been made since 1960, have not found a unique method for determining the time of technological operations, so different methods based on different scientific researches are used nowadays.

Heckner R. (1975) developed, through systematic researches, a method for calculating machine-hand times which introduces the parameter of curvature of seam and the correction of stitches sewing speed depending on the specific density of stitches. Hechner discovered that the decrease or the increase of stitches sewing speed is also affected by psychophysical abilities of workers beside the machine. When a specific density of stitch is smaller (2–3cm), due to increasing the sewing speed, it is necessary to adjust the time of machine-hand grips of sewing, because the worker reduces the stitches sewing speed unconsciously. In Table 3.6 the correction factors of stitches sewing speed are given.

Table 3.6 Correction factors of stitches sewing speed

Seam description	Type seams on cloth	Correction factor (fz)
Long, flat	Side seam of trousers	0.6-0.7
Medium long, slightly curved	Hem on the women's jackets	0.5-0.6
Short, slightly curved	Darts	0.45-0.5
Short, highly curved	Cover pocket	0.35-0.45

A practical test of methods for determining the time of sewing in real production conditions was done by Lohman (1987), who analyzed and compared five methods:

(1) Method of relationship between the speed of forming stitches and time.
(2) Method with calculated time of pressing pedals.
(3) MTM method (Deutsche MTM – Vereinigung).
(4) Conrad's method MTM.
(5) Heckman's method (sewing speed depends on the curvature and the length of seams, but also on the active reaction of workers).

Lohman thus proved that more accurate times of sewing can only be determined on the basis of average stitches sewing speed, but he did not consider the stages of acceleration and deceleration of the main shaft of sewing machines, the number of segments of joining one seam, the accuracy of joining seams and the level of training of workers.

Hopf (1978) proved, through researches and analysis of machine-hand time, that sewing time of stitches depends on the stitches sewing speed, the density of stitches, the accuracy of compiling the edges and skills of workers, applying the theory of ray lengths by MTM system.

3.5   Method with calculated time of pressing pedals

Oeser (1969) was the first one who started dealing with the problem of determining the time of sewing knitwear and he examined the maximum dynamic possibilities of sewing machines for overloch knitted seaming. He is one of the first authors who showed a diagram of dependence of stitches sewing speed upon the time in garment technology (Figure 3.6). He defined the occurrence of the acceleration main shaft, the work up to the achieved fixed stitches speed (by then it had only been a maximum sewing speed of stitches) and the deceleration of the main shaft.

Oeser indirectly pointed out, by calculating the level of efficiency, that the dynamic characteristics of the sewing machines depend on the moment of inertia of all moving mechanisms of sewing machines, the moment of inertia of the transmission system and electric motor, considering the phases of acceleration and braking.

Möller (1986) was the one who improved Heckman's method and who, apart from the stitches speed sewing, introduced the skills and abilities of workers to respond and the time for pressing the pedal while starting and stopping of electric motors into mathematical formulas for determining the machine-hand time.

Krowatschek and Ludermann (1974) did some researches about the mechanical and dynamic characteristics of the sewing machine. They developed a very modern equipment at that time for measuring the moment of inertia of sewing machine and electric motors, a device for measuring stitches sewing speed and the number of stitches in the seam by using oscilloscope with light paper. That is

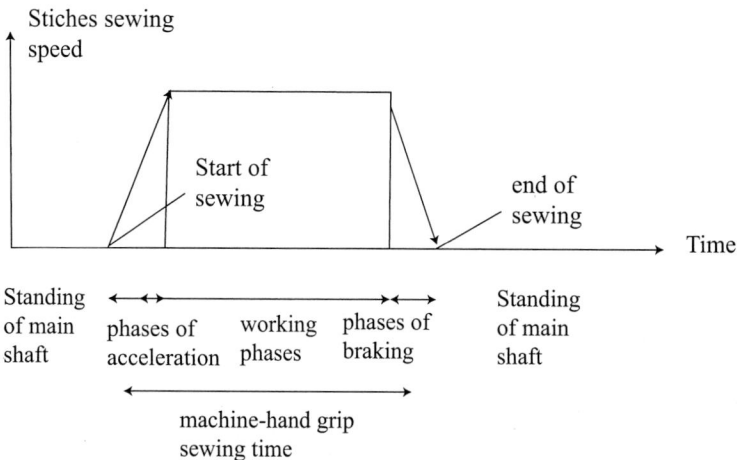

3.6 Oeser's diagram of sewing machine with the phases of acceleration and braking.

how they proved that short seams do not need sewing machines working at high speed because their speed cannot be fully expressed in the short seam.

3.6   Other methods

One of the important elements of time study and modern researches of norms is the analysis of losses of working time. Losses in work can not be avoided, and therefore it is necessary to determine, classify into reasonable and unreasonable ones and reduce them to the least possible measure by organizational or technological-technical measures. The analysis of losses first requires observing the work place according to the level of organization.

There are three types of production work places are as follows:

(1)   Open workplace – where the worker leaves the workplace during work to take work objects and resources. It is suitable for low-level organizations as well as individual production workplaces. Here, the worker loses a lot of production work time, work productivity is reduced, the machine remains unused and the production cycle is longer, so the involvement of resources in production is bigger. Researches showed that an employee works 4 hours effectively, everything else is a useless work. The manager prepares jobs for workers, and it takes him just an hour or two to do the job of an instructor.

The structure of available time for open workplace is shown in Figure 3.7, where:

G_1 [h], [%] – interruption in work because of wrong organization (no materials, machine failure, etc.),

G_2 [h], [%] – the work of employees as a result of wrong organization (wrong planning and division of work),

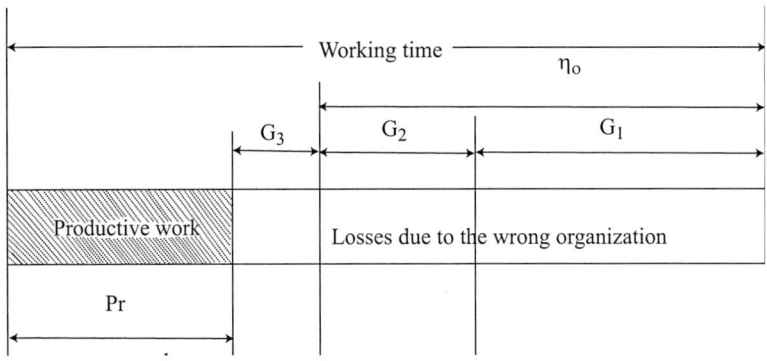

3.7 Structure of available time for open workplace.

G_3 [h], [%] – the effect is not realized because of wrong organization (wrong method of work),

Z [h], [%] – the involvement of employees (part of a total available time in which the employee works),

η_k [h], [%] – the degree of utilization of machine capacities in the workplace,

η_o [h], [%] – the degree of openness of workplace (part which remains untapped due to wrong organization),

P_r [h], [%] – the effect or productivity of workplace.

Thus, the absorption of workers in the open workplace is

$$Z^\circ = \frac{RV - G_1}{RV} 100 \ [\%] \qquad [3.21]$$

Where: RV – the available working time (e.g. 7 or 8 hours)

The degree of utilization of machine capacities in the open workplace:

$$\eta_{k^\circ} = \frac{(RV - (G_1 + G_2))}{RV} 100 [\%] \qquad [3.22]$$

Degree of openness of the workplace:

$$\eta_{o^\circ} = \frac{G_1 + G_2}{RV} 100 \ [\%] \qquad [3.23]$$

The productivity of the open workplace:

$$P_{r^\circ} = \frac{(RV - (G_1 + G_2 + G_3))}{RV} 100 \ [\%] \qquad [3.24]$$

(2) Closed workplace – where production is well-prepared. Workers are beside the machine, and work objects and resources are brought by other workers. This type of workplace is a characteristic of serial and mass production. Basic characteristics of this workplace are: working time is completely used for performing of operations so work productivity is increased, as well as specialization of workers and technical division of work; production cycle is shortened and means of work are less engaged; the degree of utilization of machines is bigger, so is the case with profitability. The employee works seven-eight hours effectively, as well as his supervisor. The structure of available time of the closed workplace is shown on Figure 3.8.

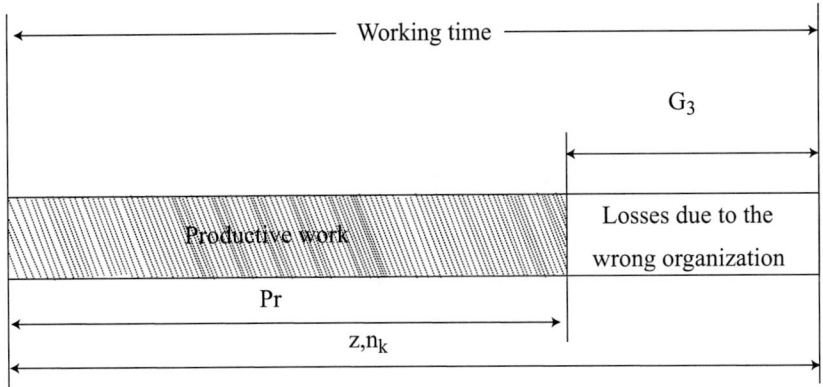

3.8 Structure of available time for closed workplace.

Involvement of workers in the closed workplace is

$$Z^z = \eta_{kz} = 100 \ [\%],$$
[3.25]

and productivity:

$$P_{rz} = \frac{RV - G_3}{RV} 100 \ [\%]$$
[3.26]

(3) Stabilized workplace – a complete rationalization and the economy of work on the workplace was made, as well as the mechanization of work and the division into the smallest operations. The employee does the same work for six instead of eight hours, because his work is easier and simplified. Apart from the job of an instructor, the manager finds new ways to simplify the work of employees. The stabilized workplace (Figure 3.9) is a characteristic of large series and mass production.

The involvement of employees in the stabilized workplace:

$$Z^s = \eta_{ks} = Pr^s = 100 \ [\%]$$
[3.27]

Degree of openness of the workplace:

$$\eta_{oz} = 0 \ [\%] \text{ because: } G_1 = G_2 = G_3 = 0 \ [\text{hour}]$$
[3.28]

In the apparel production, workplaces are open workplaces with the lowest degree of organization where the employee often stops working and leaves the workplace, so the most important causes of losses of working time are

(a) organizational-technical causes (insufficient coordination, irregular flow of material, improper scheduling of workplaces, lack of materials or

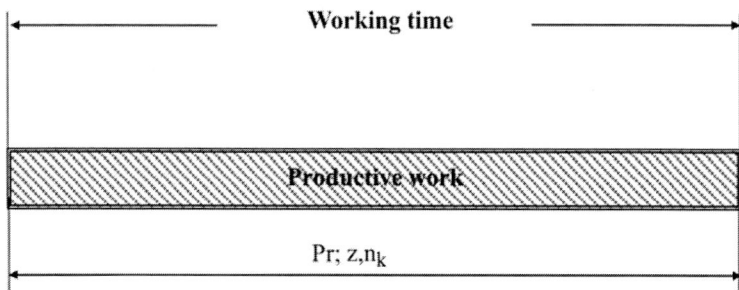

3.9 Structure of available time for stabilized workplace.

low quality materials, staff incompetence, lack of information, irrational movements while working, inappropriate division of work, overloaded workplaces, inadequate or defective internal transportation, the lack of working area, non-discipline, inadequate control while working, inadequate categorization or evaluation of work, the lack of tools or inappropriate tools, old or broken machines, etc.).

(b) organizational-biological causes (inadequate room temperature, poor lighting, inadequate ventilation and air-conditioning, inappropriate vacation schedule, unsuitable working furniture, inadequate hygiene at work, too big a distance between the rooms, etc.).

In order to solve this problem it is necessary to determine the losses and set the higher organizational level, such as stabilized or closed workplace, where working is most effective and humane. Aiming to transform the open workplace into the closed one, it is necessary to perform a range of activities in order to eliminate losses (Bulat and Bojkovic, 1999):

(1) Programming of production.
(2) Profit analysis of products.
(3) Selection of production programmes.
(4) Determining the optimal production plan.
(5) Calculating the optimal size and number of series of products from the optimal plan.
(6) Choosing the optimal types of resources (tools, machines) following the production programme.
(7) Determining the required number of resources per types.
(8) Determining the required number of workers per shifts and the professions.
(9) Determining the required number of workplace per types.
(10) Determining the optimal schedule of work.
(11) Optimal division of work.
(12) Determining the optimum order of performing work tasks.

(13) Scheduling production and launching of work tasks.
(14) Supply of workplaces with jobs.
(15) Study, measurement, improvement and humanization of the workplace, etc.

In order to determine the loss of working time, two methods are used:

(1) Picture of working day is a simple method of recording losses in one day of at least three employees with a stopwatch timer.
(2) The method of current observation (Work sampling, Ratio delay method, Méthode des observations instantanées) – recording the losses of workplaces randomly chosen, and counting the number of employees working at that moment. This statistical method of observation of phenomena on the basis of their frequency was first applied by the British statistician LNC Tippett in 1934 in the textile industry, and it has been mostly applied ever since.

Work sampling is used for:

(1) Determining the degree of capacity utilization.
(2) Determining the degree of openness of the workplace, groups of workplaces and organizational units.
(3) Calculating the level of organization of organizational units.
(4) Determining the preliminary–final time, help-time and additional component time within the calculation of the normal production time.

The process of implementation of method of current observation by Martinovic (2002) is as follows:

(1) Determine the aim of recording, the aim of applying methods.
(2) Inform the people who work in workplaces that will be recorded about the aim and the method of recording.
(3) Make a schematic view of objects that will be recorded.
(4) Define the path of a recorder, determine the best position of the recorder for each object to be recorded and draw a schematic view.
(5) Create forms (document) for recording.
(6) Train the employees who will record.
(7) Determine the time of a starting point of each recorder by using a table of random numbers.
(8) Adopt the accuracy of the indicators that should be determined by the work sampling.
(9) Calculate the required number of notes and tours.
(10) Control whether the process of recording is performed normally. If not, stop the recording.
(11) Sort out the recorded material.

(12) Calculate indicators (parameters, values, size, etc.) and determine their accuracy.
(13) Undertake analysis of the results obtained.
(14) Make appropriate conclusions.
(15) Take the necessary measures.

After you complete the recording of data by the method of current observation, the analysis of recorded data is performed, i.e. (Sajfert and Nikolic, 2002):

(1) recording by elements,
(2) developing circular diagrams,
(3) calculating of capacity utilization,

$$q = \frac{\text{number of work notice}}{\text{total number of notice}} \cdot 100 \ [\%] \qquad [3.28]$$

(4) calculating the losses in work,

$$p = \frac{\text{number of non work notice}}{\text{total number of notice}} \cdot 100 \ [\%] \qquad [3.29]$$

(5) calculating the additional coefficients

$$K_d = \frac{\Sigma G_p}{\Sigma R + \Sigma G_{NP} + \Sigma G_{NE}} \qquad [3.30]$$

Where: G_p – recognized losses in the working norm,
R – number of notes,
G_{NP} – unrecognized losses,
G_{NE} – losses because of indiscipline.

In order to test the accuracy of indicators of capacity utilization degree, this formula is used:

$$t = 100 \cdot \left[1 - 2\sqrt{\frac{l-q}{q \cdot n}} \right] \qquad [3.31]$$

Where: n – total number of notes
As an example of method of current observation, Table 3.7 shows the observed document obtained in a sewing room during a working week (six working days).

Table 3.7 Sheet for recording of working time losses

Date	Time of rounds	Direct work	Prepare machine for operation	No job	The machine does not work	Short justified delay	Unjusti- fied delay	No material, tools, docu- ments	Not work because of quality control	The worker is not in the work place	Σ Losses
15 September 2008.	7h10	||||||||	III	I		III	I	II	I	I	12
	8h10	|||||||||||	I		II	I	III		II		9
	9h10	|||||||||||||		I			I			I	3
	10h10	||||||||	III		I	III		I	II		10
	11h10	|||||||||||		II			III			II	7
	12h10	|||||||||||	I		I	I					3
	13h10	|||||||||||||		I			I		IIII	I	7
	13h45	|||||||||||||||									/
16 September 2008.	7h10	|||||||||	IIII			IIII		II			10
	8h10	||||||||	III		I		IIII	I	I	III	12
	9h10	|||||||||||||||									/
	10h10	|||||||||||	IIII			IIII					8
	11h10	|||||||||		I	II		IIII		II	I	10
	12h10	|||||||||||	III			III		I			7
	13h10	||||||||||		II			III		II	II	9
	13h45	|||||||||||||||									/
17 September 2008.	7h10	||||||||||	II		II	III			III		10
	8h10	|||||||||||		II			II	II		II	8
	9h10	||||||||||	III			III			II		8
	10h10	|||||||||||||					III	I			4
	11h10	||||||||	III		III	III			III		12
	12h10	|||||||||	I	II		I	III			II	9
	13h10	|||||||||||||||					I				1
	13h45	||||||||||||||									/
18 September 2008.	7h10	||||||	II	II	II	II		I	III	II	14
	8h10	|||||||||||||					II	II			4
	9h10	|||||||||||||		II						II	4
	10h10	||||||||||	III		II	II			III		10
	11h10	|||||||||	I	II		I	III	I		II	10
	12h10	|||||||||||||	I			I	I		II		5
	13h10	|||||||||||||		II			I	I		II	6
	13h45	|||||||||||||||									/
19 September 2008	7h10	||||||||||||||	I	I		I				I	4
	8h10	|||||||||||||||			I		I		I		3
	9h10	|||||||||||||||	I					I			2
	10h10	|||||||||||	II	I		I	I		II	I	8
	11h10	|||||||||||||||			I		I	I			3
	12h10	||||||||	III	II		I			III	II	11
	13h10	|||||||||||			I		II	II			5

(Continued)

20 September 2008.	Time										
	7h10	IIIIIIIIII	II	I		I	II			I ı	7
	8h10	IIIIIIIIIII						I	I		2
	9h10	IIIIIIIIIII		I							1
	10h10	IIIIIIIIIII	II				I			II	5
	11h10	IIIIIIIIIII						I			1
	12h10	IIIIIIIIIII		I		I					2
	13h10	IIIIIIIIIII									/
	13h45	IIIIIIIIIII									/
	Σ	734	46	30	18	40	45	20	37	30	1000

After the recording of data by Method of current observation has been completed, the analysis of recorded data is performed (Table 3.8).

On the basis of data from Table 3.8, the level of capacity (q) is calculated, as well as losses in work (p) and an additional coefficient (K_d). The level of capacity to [3.28] is $q = \dfrac{780}{1000} \cdot 100 = 78\%$.

Table 3.8 Data obtained by work sampling

Number	Description	Number of notice	%
1.	Direct work	734	73.4
2.	Prepare machine for operation	46	4.60
3.	No job	30	3.00
4.	The machine does not work	18	1.80
5.	Short justified delay	40	4.00
6.	Unjustified delay	45	4.50
7.	No material, tools, documents	20	2.00
8.	No work because of quality control	37	3.70
9.	The worker is not in the workplace	30	3.00
	Total:	**1000**	100

Losses in work [3.29] as $p = \dfrac{220}{1000} \cdot 1000 = 22\%$

Calculating the additional coefficient by [3.30]:

$$k_d = \frac{\Sigma G_P}{\Sigma R + \Sigma G_{NP} + \Sigma G_{NE}} = \frac{30 + 18 + 40 + 20 + 37}{780 + 30 + 45 +} = 0.169$$

The accuracy of indicators of capacity utilization over the formula [3.31] is

$$t = 100 \left[1 - 2\sqrt{\frac{1 - 0.78}{0.78 - 1000}} \right] = 96.64\%, \text{ i.e. with 1000 notes the uncertainty is}$$

3.36%.

Considering the fact that the machine delay $Z = \frac{220}{1000} 100 = 22\%$, then the

reliability is t = 88.1%. In order to make the accuracy of machine delay 95%, it is necessary to calculate the number of notes:

$$n \geq \frac{4.10^4}{5^2} \cdot \frac{1 - 0.22}{0.22}$$
$$n \geq 1596$$

According to this, 1000 notes were sufficient for capacity utilization q, whereas the machine delay required 1596 notes.

Testing the accuracy of other indicators:

➢ direct the work t = 96.19% (3.81% uncertainty),
➢ preparing machines for work t = 71.19% (28.8% uncertainty),
➢ no job t = 64% (uncertainty 36%),
➢ machine failure t = 53.28% (46.72% uncertainty),
➢ short justified delay t = 69% (31% uncertainty),
➢ unjustified delay t = 70.8% (29.2% uncertainty),
➢ no materials, tools and documents t = 55.72% (44.28% uncertainty),
➢ no work because of quality control t = 67.7% (32.3% uncertainty),
➢ employee is not in the workplace t = 64% (uncertainty 36%).

Method of current observation can also be used as a method for determining the normal production time, if one operation is performed on many workplaces. Working time, break time, time for physiological needs, acknowledged delays etc. During the application of method of current observation a time of recording is established simultaneously, as well as the number of pieces produced and the mean value of rhythm of workers' performance. Obtained values in percents are converted into the corresponding times in minutes, according to the formula:

$$t_{iz} = \frac{\text{total time} \cdot \text{perecent of work} \cdot \text{mean value of rhythm}}{\text{total number of opperation pieces produced}} \qquad [3.32]$$

There are many advantages and disadvantages of method of current observation which can be defined on the basis of this example. The advantages are as follows: one recorder can record a number of workplaces; less work and less costs than in continuous recording; the period of observation can be extended to avoid periodical variations; workers, in time, relieve themselves of feeling continuously observed, and they have a normal attitude to work – more accurate data; it takes less time for the recorder to prepare for recording; computer data processing can be used. Disadvantages of method of current observation are as follows: non-economical for applying in one workplace; not getting detailed information; and production workers find it hard to understand.

Reducing the number of workers in manufacturing, automation, and increasing the number of employees in administration will make this method of current observation, as a method for determining the normal production time, widely used in the future.

References

1. Adamović Z and Alihodzic A (2002). *Upravljanje proizvodnjom*, Zavod za udzbenike i nastavna sredstva, Srpsko Sarajevo.
2. Barnes RM (1947). *Motion and Time Study*, J. Wiley & sons, New York, Chapman & Hall limited, London.
3. Barnes RM (1968). *Motion and Time Study: Design and Measurement of Work*, Wiley, New York.
4. Bulat V and Bojkovic R (1999). *Organizacija proizvodnje*, ICIM, Krusevac.
5. Colovic G and Petrovic V (2001). 'The analysis of working time losses in a technological process of the production of men's T- shirts', *1ˢᵗ International Ergonomics conference, Energonomy 2001*, Zagreb, 143–153.
6. Colovic G, Paunovic D and Savanovic G (2009). 'Analysis of Classical and Modern Production Line for Production of Male Denim Jacket', *International Scientific Conference UNITECH 09*, Gabrovo, s9–p95.
7. Colovic G, Petrovic V, Djordjevic J and Paunovic D (2003). 'Analiza metoda za odredivanje vremena izrade odevnih predmeta', *V simpozijum Savremene tehnologije i privredni razvoj,* Tehnološki fakultet, Leskovac, p. 149.
8. Heckner R (1975). 'Zeitwertkartei und prozeßzeitberechnung für die Näherei', *Bekleidung+Wäsche* 27, 18, pp. 1081–1084.
9. Hopf H (1978). 'Benötigen wir ein neues Unterweisungsprogramm für Nähereinen', *DOB + haka praxis* 13, 7, pp. 456–461.
10. Krowatschek F and Ludermann (1974). 'Der Naehvorgang als Kybernetisches System', *Band 5*, Forschnunggemeinschaft, Bekleidungsindustrie R.V. Berlin, pp. 82–89.
11. Krowatschek F and Nestler R (1994). 'Bessere Nähguttransport durch Langsamnähen', *Bekleidung /weare* 36, 17, pp. 13–15, 37–39.
12. Lochmann G (1987). 'Über di Berechnung von Prozesszeiten bei Nähmaschine', *Bekleidung+Wäsche* 37, 11, pp. 15–19.
13. Martinovic M (2002). *Organizacija proizvodnje*, Visa tehnicka skola, Uzice.
14. Martinovic M and Colovic G (2007). 'System PPORF in garment industry', *Serbian Journal of Management*, 2 (1), pp. 73–81.

15. Maynard HB, Stegemerten GJ and Schwab JL (1948). *Methods-time Measurement,* McGraw-Hill, New York.
16. Moll P (2000). *New Courage for Reducting the Personal Cost,* International Symposium AVANTEX, Frankfurt, pp. 676–677.
17. Möller W (1986). 'Realistische Prozesszeit für das Maschinennähen', Bekleidungstechnik 36, 17, pp. 30–32.
18. MTM (1984). Standard Data Fertigunsbereich Maschinennähen.
19. Oeser W (1969). 'Ermittlung von Normativen für Nahtlänge in Abhängigkeit von der Maschinengeschwindigkeit', *Bekleidung und Mashen-weare* 8, 3, pp. 104–106.
20. Rogale D, Ujevic D, Rogale S F and Hrastinski M (2000). *Tehnologija proizvodnje odece sa studijom rada,* Tehnološki fakultet, Bihac.
21. Sajfert Z and Nikolic M (2002). *Proizvodno poslovni sistemi,* Tehnicki fakultet M. Pupin, Zrenjanin.
22. Taborsak D (1971). *Studij rada,* Tehnicka knjiga, Zagreb.

4

Ergonomic workplace

Abstract: Proper ergonomic design of each workplace together with finding a suitable method of work with the appropriate time standards ensures a better structure of technological operation with the increased efficiency of sewing machines. Position for working at a sewing machine should enable the mobility of limbs, ergonomically favourable arrangement of working zones and the visible ones and a stable balance state in performing the work process.

Keywords: ergonomics, workplace, movement analysis.

4.1　Ergonomic workplace

Employees in industry often deal with problems of health, safety, motivation and efficiency of workers, depending on the production process and working space. According to the European Foundation for the Improvement of Living and Working Conditions in EU countries 62% of workplaces include repetitive movements of arms and hands, 46% body positions that cause pain and fatigue, 35% handling heavy burdens, 56% of jobs require work with the short timescale and 54% work in a fast pace, 42% do not allow vacations and 31% are against free rhythm. About 40% of all employees during at least 25% of their working time are exposed to at least three factors for the development of diseases of systems for movement. In order to solve these problems, it is necessary to take steps to prevent or mitigate the troubles arising in the workplace during a production process which also have a bad effect on the life and work of workers in industry. Basic steps are as follows:

(1) *Technical–organizational* – the introduction of a production technology often makes a serious problem in psychophysical strain for a man and his safety at work, social interaction and communication, the monotony which occurs due to depletion of work and fatigue due to the hardship and rhythm of work. In order to solve these problems, it is necessary to take technical–organizational steps such as separating humans from machines, automation, replacement of workplace and expansion of work.

(2) Technical–safety – to prevent accidents and troubles at work, they must be taken care of when introducing new machines, tools and designing workplace.

(3) Ergonomic – when designing the workspace, machines, tools and devices, to facilitate human work and adjust the work to his psychophysical abilities. Ergonomic measures include the problems of methods of work.

Ergonomics (Greek: Ergon = work + Nomos = custom, law) is an interdisciplinary field of science about the work that deals with the research of biological, psychological and sociological moments of human labour in the adaptation of human to machine and vice versa. The term ergonomics was created by Murrell K F H in 1949. Human Factors Engineering and Human Factors is a synonym for ergonomics.

There are several definitions of ergonomics:

- According to the "father of ergonomics," Alphonse Chapanis (1985) ergonomics is a discipline that examines the characteristics of behaviour, abilities, limitations and other characteristics of people and applies discovered information to the design of tools, machines, systems, tasks, jobs and environment so that they can be used productively, safely, comfortably and effectively.

- By the International Ergonomics Association, ergonomics (or human factors) is the scientific discipline concerned with the understanding of interactions among humans and other elements of a system, and the profession that applies theory, principles, data and methods to design in order to optimize human well-being and overall system performance. Ergonomics contribute to the design and evaluation of tasks, jobs, products, environments and systems in order to make them compatible with the needs, abilities and limitations of people.

- By James H. Stramler (1993) ergonomics is that field which is involved in conducting research regarding human psychological, social, physical, and biological characteristics, maintaining the information obtained from that research, and working to apply that information with respect to the design, operation, or use of products or systems for optimizing human performance, health, safety, and/or habitability.

- Ergonomics is now predominantly observed as an interdisciplinary part of science of work. According to R. Hackstein, the science of work is "a combination of theoretical, descriptive and experimental, natural and social sciences...about human labour as a conscious and planned, body and spiritual activity which aims to satisfy basic needs first, and then the other ones..."

Put simply, the science of work deals with parsing and designing of working systems and working environment, trying to establish, on the basis of scientific knowledge, all necessary measures that would improve and facilitate the work and life of a man in industry. The main difference between the ergonomics and science of work is that the area of ergonomics is theoretical, and ergonomics should be viewed in its practical dimension – as technology.

The task of ergonomics is to optimize the human–machine–environment system, adjusting the working conditions to physical, physiological and psychophysical nature of man, considering relevant differences among people in regard to jobs and the workplace. Thus, the task of ergonomics is the alignment of parts of the system (man–machine–environment–organization) with the help of component parts of the science of work, namely anthropometry, work physiology, psychology, sociology, work technology, work pedagogy and organization of work. Not only does ergonomics increase efficiency and productivity of manufacturing or business systems, but it also improves the health, safety and comfort of man in his working environment.

German School of ergonomics defines macro-ergonomics more orientated to sociology, and micro-ergonomics orientated to improve performances of whole system and reduces stress caused by work.

The main goal of micro-ergonomics is to improve the performance of the work system and reduce stress analysis: task, working environment and man–machine interaction. The concept of stress–strain is the traditional approach to assessing the working system. The basic concept is that every workplace is characterized by external factors, which are the same for all individuals who react differently depending on the individual characteristics and abilities. The stress is different from the parameters of stress (defined by numbers), stress factors (given only descriptively) and the time of stress exposure. In order to understand the factors that have the influence on work, the structure of Human–Machine System (HMS) must be examined by monitoring human labour in relation to the information and the speed of information flow. This includes setting up the task, its putting into action, the performing of the task and the result of performing the task. A feedback closes the control loop formed for the human–machine system and shows that the operator is generally capable of comparing the task and the result. All the disorders of this process are the impacts of environment. While analysing the task defining, there are

- tasks with predominantly physical strain. Here, some people define the difference between static and dynamic physical work. In both cases, stress can be quantified by defining the physical requirements.
- tasks with predominantly mental strain (intellectual work). A generalized concept for defining this stress in numbers does not exist. Intellectual work is therefore taken as a factor of stress.
- tasks with both requirements (physical and mental).

While analysing environmental impacts (ergonomics of environment), there are

(a) physical environmental impacts which can be measured, as well as their impact on man and can be assessed quantitatively (lighting, noise, mechanical vibrations, air-conditioning, toxic gases, radiation, dust, dirt and humidity).

(b) social environmental impacts which can not be measured physically and therefore are marked differently (they are sometimes called a work sociology or industrial psychology).

Macro-ergonomics deals with the systematic structure and organization of workflow considering the task, the content and the time factors. It can be divided into the organizational structure and the organization of work processes. The goal of macro-ergonomics is not an individual workplace but the interaction between many workplaces. Its goal is testing ergonomic requirements at this level. In this context, the term macro work can be used as well. The unit area of macro-organization can be divided if a unit operating system is transferred to a higher operating task in the context of a group. The analysis of workflow provides the right-in-time information necessary for the task that should be performed within the organizational unit and on the basis of internal dependences. This allows the specifying of capacity requirements of humans, the means of production and the time of their utilisation. In particular, ways of communication and possible losses of information are determined in order to optimize the interaction between workers and working funds. The development of innovative telecommunications and computer technology is a new challenge for the organization of work.

Environmental conditions are defined as those that do not directly influence the work process and the process of communication but, similarly to micro-ergonomics, change it indirectly. It is important to emphasize the difference between the impact that can not be changed by the organization of work and the influence that can be optimized by the corresponding design of workflow and organizational structure of cooperation.

On the output side of the working process, performance improvement can be obtained by organizational steps, with personally verified methods of the quality of working object and the results respectively developed on one side, and on the other there is the worker under the influence of motivational factors.

Many of the cases described can not simply be proved by an experiment. Therefore, partial detailed simulation methods are often used, and through cumulative key numbers they describe personal acceptance, personal qualifications, waiting time, overlapping by simultaneous upcoming order to provide the assessment of organizational changes and new structures.

4.2   Division of ergonomics

There are several divisions of ergonomics according to the types and specific human traits and characteristics of human interaction with the environment.

Types of ergonomics are conceptual ergonomics, system ergonomics, corrective ergonomics, software ergonomics and hardware ergonomics.

4.2.1   Conceptual ergonomics

Conceptual ergonomics deals with the designing of ergonomic measures in the very beginning of construction of a working system and therefore is the cheapest one. This ergonomics includes the tasks of improving the conditions of life and work in two areas:

- the area of humanity and
- the area of economy.

In the area of humanity, the ergonomics must reduce the strain of workers, reduce the risks at work, allow a holiday, increase the satisfaction and interest in work and make work pleasant. It is important to reduce health problems at work, improve the protection while working, reduce the harmful influence of environment and perform the work easily.

The tasks of ergonomics from the perspective of economy are increase the accuracy of work, speed up the rhythm, ensure the feasibility of labour, reduce the work requirements, reduce costs, facilitate decision-making, improve information flow and utilization of time.

Ergonomics must provide the increase of motivation, quantity and quality of work. In order to meet the specified requirements, they must shape the ergonomic measures that arise as a general result of a system ergonomics.

4.2.2   System ergonomics

The task of system ergonomics is to take care of the coordination functions of a manufacturing system. It takes care of personal and mechanical functions in which the man in the production system must not be under any kind of strain. System ergonomics pays attention not only to some parts of the system (man, machine, environment), but also to the whole system. According to B. Doring (1976), system ergonomics has several areas of interest:

- ➤ designing organization of working system,
- ➤ organization flow (process) of work system,
- ➤ designing workplace,
- ➤ designing working areas,
- ➤ designing working environment and
- ➤ selection and training of employees.

The base of a system ergonomics is a conceptual ergonomics. After the situation is established conceptually, the system ergonomics decides about the steps to be taken. System ergonomics is a sort of methodical, technological procedure that is performed when developing a workplace. During the implementation of the functions of system ergonomics, human psychophysical capabilities must always be taken care of.

4.2.3 Corrective ergonomics

Corrective ergonomics occurs in the later period of realization or utilisation of the working system. It implies meeting the ergonomic requirements subsequently, so it is less successful and more expensive than the previously mentioned types of ergonomics. It is subjected to many limitations because the ergonomic principles are neglected in its development phase. Corrective measures are therefore based on reliable experiences. Corrective ergonomics is based on the principle: "You see – you look – you improve". The representative of this ergonomics is Barnes (1968), who defined 22 principles for the rationalization of work as a whole and they are 8 principles related to the economy of movement, 8 principles related to the regulation of workplace and 6 principles related to methods and principles of designing tools and equipment.

4.2.4 Software ergonomics

Software ergonomics is an interdisciplinary part of science of work that deals with direct or indirect effects of software products in the human–machine–environment system. It includes biological, psychological and social aspects of interaction between a man and a software. The aims of software ergonomics are

- improvement of adopting information technology,
- improvement of work motivation,
- increase of work competence,
- development of personality and
- optimization of strain in introducing new technologies.

By introducing computers, a man no longer manages the machine directly but indirectly, because of which he must dispose of all the components that will allow him a certain level of freedom when coping with work tasks and goals within the cores of software ergonomics.

The usage of information technology allows the increase of production with the help of new technologies; it enables the increase of efficiency of processing information by introducing better methods and procedures.

Software ergonomics is also concerned that because of the strain transfer from physical to mental side of man, a worker shouldn't be under too big or too small strain. That's why, while working at the computer (a sitting workplace), it is important to know the following:

(1) Feet must be laid flat on the floor.
(2) The knee joint must be bent to 90°, i.e. set from 20° to 45° below the hips.
(3) Elbows must be laid parallel on the table or with minor deviations.
(4) Shoulders must be flat and straight, but also relaxed.
(5) The screen must be at least 50 cm away from the head and must be in the eye sight without moving the head.
(6) Wrists must be flat and under no strain.
(7) Keys on the keyboard must be pressed very lightly without using much force.

In the countries with developed industry the concern about adjusting the workplace to the worker gradually becomes a subject of interest because of physiological (health), psychological (selection and satisfaction), and economic reasons (production quantity and quality). The ergonomic position while working on a computer, according to the European Union Council Directives 90/270/EEC, is shown in Figure 4.1.

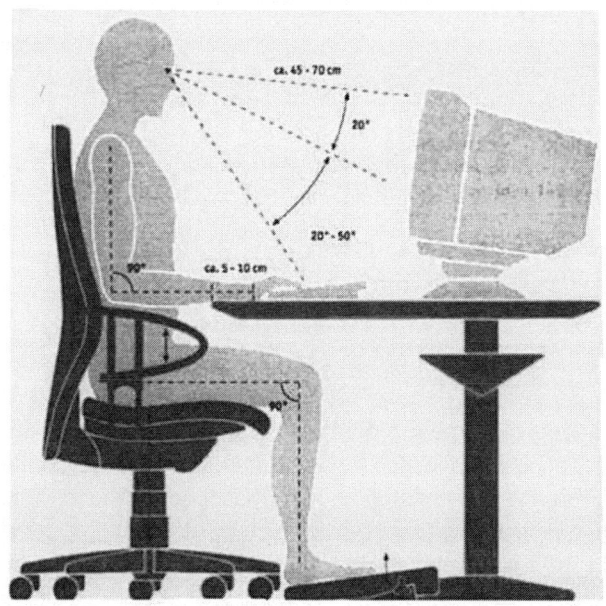

4.1 Ergonomic positions while working on computer.

4.2.5   Hardware ergonomics

"Classical ergonomics" actually means this type of ergonomics. It does not deal with the working content such as technical–physical components of computer systems, direct and indirect environment of system; for example, suitable construction of the place of the apparatus, the height of the place, a chair, its parameters and reflective surfaces.

According to the specific human traits and characteristics of human interaction with the environment, ergonomics is divided into: physical ergonomics, cognitive ergonomics and organizational ergonomics.

Physical ergonomics deals with anatomical, anthropometrical, psychological and biomechanical characteristics of human beings in their relationship with physical activity attitudes towards work, handling with materials, frequent injuries due to movement, muscle–bone disorders, organization of working space, safety and health.

Cognitive ergonomics deals with mental processes such as perception, memory, thinking and mobility and the way they are affected by the interaction with the remains of the observed system. The most important aspects include mental effort, decision making, interaction with computers, human reliability and work stress.

Organizational ergonomics studies the optimization of socio-technical systems, including their organizational structure, rules and processes. This ergonomics includes communication, organization of work, teams and teamwork, communal ergonomics, cooperative work and management.

4.3   Ergonomic conditions

Designing a workplace is a very important segment of the intangible strategies of motivation since the attitudes towards work and the pleasure of it significantly affect the motivation at work, and also the entire life of the individual. Programs of redesigning the workplace try very hard to make the job interesting and challenging. Significant individual approaches to designing workplaces are job rotation, where people are periodically moved from one specialized job to another in order to prevent monotony and boredom. However, the real motivational potentials are best activated by the enrichment of work which spreads vertically, including various tasks and skills, responsibilities and autonomies.

Basic characteristics of the work to be taken into account when designing the jobs are skill variety, identity and task integrity, the importance of the task, autonomy and feedback. In addition to the individual, there are also group accesses to designing a workplace. Thus, in integrated working groups workers are given a number of task assignments instead of one, and in autonomous working groups they are only given the goal, and it is up to them to determine the working responsibilities, time for rest, etc.

Participation as the level of taking part of employees in decision-making processes concerning important aspects of work has a significant influence on the increase of employees' motivation, encouraging creative and overall potentials of people, improving the quality of decisions and the success of BPS.

Management by Objectives is an important strategy of modern management in raising motivation, quality, human resources, flexibility and responsiveness to changes in the environment. This defines areas of responsibilities and standards of behaviour for each production unit.

One of the biggest problems is the resistance of the workers who often do not believe that the system of stimulating rewards is objective and fair. One of the important conditions of success of stimulating rewards is to gain the confidence of workers, so salaries should be supplemented with good designing of a workplace.

In order to design a workplace, it is necessary to know ergonomic conditions. Ergonomic conditions are physiological, psycho-sociological, anthropometric and ecological conditions of work.

(1) Physiological conditions are related to studying working capacities of man and various influences on working capacity. Physiology is the study of functioning of certain organs of living beings, as well as chemical and physical processes that are being carried out inside of them. Physiology of work is a special branch of physiology which is limited to the body of a man who works. A man is most capable of making best sport (and physical) results when he is about 25, and of mental (and organizational) results when he is about 45 years old.

   Rational usage of working capacity of man is not only a matter of being humane, but also of being economical. If a man in his early age or when too old is entrusted tasks to which he is not physiologically (physically) equal to, it significantly affects the efficiency of human life. Such a man is ill much more during his working life.

(2) Psycho-sociological conditions of work are related primarily to achieving a sense of satisfaction in work. The satisfaction of a man who works is influenced by many factors which are largely related to the motivation for work. Mental fatigue occurs due to excessive mental strain during a certain period of time. It is manifested as fatigue which can be eliminated by frequent periods of rest during shifts. Monotony in work has a negative influence not only on the concentration of workers, but also on his satisfaction. It can be eliminated by changing the organization of work. Mental saturation is a condition which occurs when there is a resistance against accepting or continuing the

performing of a particular work activity. It is manifested as irritability, anxiety and a loss of will for work. Stress occurs when a worker feels threatened due to strain or danger to which he is exposed in the course of performing the work. Stress occurs when a worker believes that he is under too much strain, or when he has the impression that he can not affect the situation.

(3) Anthropometric working conditions refer to the conditions in which a man in the workplace should be ensured in order to adjust the work performed up to anthropometric (dimensional) features of man.

(4) Ecological conditions of work are a new attitude towards the environment.

Unless ergonomic principles are complied with, a man is exposed to a series of risk factors which has been confirmed in thousands of epidemiological researches, laboratory tests and histories of diseases including action force, repetitive movements, uncomfortable body position, bad posture, vibration, stress and coolness. Some examples of risk factors that may lead to the occurrence of musculoskeletal and other disorders are as follows:

- *Unnatural and static positions.* Bending or lowering the body due to holding or lifting heavy objects; pulling out or pushing objects into blocked areas; frequently repeated tasks which include leaning, bending forward, kneeling or squatting, working with arms bent or distorted, using hands below waist or above the shoulders; standing or sitting most of the shifts; working with arms or hands in the same position over a longer period of time without changing posture or resting.
- *Putting a lot of force into the movement.* Lifting (lifting heavy loads with one hand or without the help of mechanical devices; lifting heavy loads with bending, reaching above shoulders or leaning) pushing, pulling, carrying (manual cranes for pallets or other carts which are difficult to move; uneven surfaces and cracks in the floor or panels with low edges that can catch wheels while pushing; pulling objects instead of pushing them; manual transport of heavy objects to big distances) and using tools which are too small or too large for the hands of workers.
- *Repetitive movements* (rapid movements of hands; movements that are performed for several hours without rest; jobs that require a repeating force of the fingers – packaging, putting labels on products).
- *Contact stress* (contact with sharp or hard edges, working with machines for cutting – knives).
- *Vibrations* (using tools on the electric drive – vibrations of hands and arms, driving forklifts, trucks and other vehicles – whole body vibration).
- Coldness – working in cold environment without proper clothing.

Many of the musculoskeletal injuries can be prevented. Employers must reduce any risks identified in the process to the lowest possible level by introducing control measures. It is important to consult the workers then because they know most about the job. This means that the job should be adjusted to the worker, not the worker to the job.

Just like designing the workplace in the classical industrial situations, there are several ways of performing the work nowadays: standing, sitting, and mixed (the best) workplace. In designing the workplace it is necessary to use the ergonomic rules that will decide on the posture of the body at work, e.g. comfortable position of hands, the spatial freedom of movements of hands in all directions when standing, sitting and combined work, the height of workplace, microclimate, etc. It aims to reduce psychophysical load in order to prevent fatigue, monotony, stress or mental saturation.

In constructing a new workplace or analysing the old one it is necessary to pay attention to the following factors:

- Anthropometric factors (the height of a man because of reaching to handles or buttons on the machine).
- Biomechanical factors (the way in which a force acts on our wrists, the method of transporting material and handling it, body posture at work – sitting or standing, if there is a possibility of adjusting the height of chairs).
- Physiological factors (the way of functioning of human body).
- Factors related to the construction of work (jobs are divided into smaller segments; introducing workers to the goal and purpose of work).
- Factors related to the information and the control of work (if the information is presented in the simplest way).
- Working environment factors (noise and vibration, lighting, air conditioning, protective equipment at work).

In designing a workplace it is necessary to pay attention to the conditions of work and harmonize with four characteristics of workers. These are motor-physical (height, weight), sensory (hearing, vision), mental (intellectual ability, memory, attention) and spiritual (morality) characteristics. It is necessary to avoid unnatural body postures, such as leaning on the back or aside, and to lean forward 15° maximum; working with arms held out, because it reduces the accuracy of work; squatting and stooping. It is necessary to take into consideration the relationship between static and dynamic muscle work (the relationship between the angles of different parts of the body, the mass distribution of individual body segments, the duration of a movement and the risk of a posture) using

- indirect methods – taking photographs or recording workers,
- direct methods – watching the man working,

- subjective methods – analysis of employees, i.e. when a worker is asked about his movements at work.

4.4   Movement analysis

The study of schedule of a technological operation reveals the schedule that allows a shorter path of movement and optimal sequence of grips and movements in the operation. Better schedule and sequence lead to the increasing of labour productivity and better humanization in work, as well as better utilization of existing resources, shortening the total length of moving the objects of work, reducing the number of grips, shortening the duration of the operation.

Organizational model is different from the production technology and it can be made in a written form or acquired during a long series of repetitions, and it is caused by:

- schedule in the workplace,
- sequence of performing the task and
- interdependence of performing the grips.

The following rules are particularly important:

- workplace must have the optimal size,
- working conditions should correspond to standards,
- equipment should enable work in a standing or sitting position (employee elects),
- equipment should be located in the optimal zone is selected according to the frequency of handling,
- arrange the equipment to provide the optimal sequence of movements in the operation and
- arrange the equipment for supplying the workplace so that it should be optimal in relation to employees and inter phase transport.

When planning the performing of an operation, i.e. determining the ways of performing an operation, certain principles, which provide better performing of operations and less strain of a man at work, are very important.

- Both hands should work without interruption.

This principle avoids unnecessary delays in work, and the usage of one hand puts uneven pressure on human body.

- Hand moves should be simultaneous.

This principle ensures less mental effort, because a simultaneous beginning or ending of two movements requires only one command of human brain.

- Hand moves should be symmetrical.

The movements of left and right hand are "the same" for the human brain only if they are symmetrical. This means that each hand moves as an image in the mirror of the other hand. It takes only one mental effort, not two, to control the symmetrical movements.

- Hand moves should be reduced to the lowest class.

This principle suggests that less force and more speed require less muscle mass and vice versa. Thus, human energy will be used rationally. Classes of muscle mass correspond to classes of movements. Movements "from the elbow" are the most suitable ones for removing the product. As for the movements that must be frequently repeated, such as the wrench and wring, the optimal movements are "from the wrist".

- Use the force of inertia in movement.

When stopping the load which is moving (breaking), there is the force of inertia that pulls the load further in the direction of movement. This force depends on the speed of movement and the weight of load. This force should be used when planning to perform an operation.

- Movements should be punctured.

A lot of researches proved that the movement with sudden changes of direction causes the loss of up to 80% of time on controlling the power of inertia while stopping and developing of speed when moving in opposite directions. That is why only smooth and continuous movements are recommended.

- Movements should be ballistic.

The characteristics of a ballistic movement are that at the very beginning of the movement the hand muscle gives the force to the burden. Hand muscles are relaxed during the movement.

- Movements are to be performed in natural rhythm.

Man is prone to the rhythm in work and this tendency should be used. Natural rhythm eliminates unexpected delays. Durations of operations, which are repeated, are shorter and more equal if a natural rhythm of movement is used. In this way a man develops working habits and does not put any effort and time into making decisions during the work.

The following rules are especially taken care of:

- whenever possible eliminate the grip,
- whenever possible connect with the previous or the following grip or movement,
- whenever possible do a set of grip in several areas simultaneously,
- whenever possible release hands and perform grips by feet,

- change the sequence whenever it leads to more efficient work,
- loaded grip should be performed by those parts of the body whose features are most appropriate,
- sequence of hand movements should be designed to be simultaneous, symmetrical, and in opposite directions.

In order to study the sequence of grips and movements when performing an operation various methods are used, such as:

(1) *The method of model* – the model map – for studying the arrangement of equipment and commands when performing an operation, i.e. a technical means a model of work is drawn into, such as:
- changing of arrangement of equipments in the workplace,
- providing the procurement of organizational supplies and perform their schedule,
- reconstruction of existing equipment.

The method of model allows the increase of productivity, the improvement in humanization of work and better utilization of the surface and volume of a workplace.

(2) *The method of thread* studies the arrangement of equipment in the workplace while performing an operation. By pulling a thread from one landing point of the object of work to another along the approximate path of their movement, it is possible to study the schedule of the workplace, taking into consideration the impact of movement of the object of work. The method of thread allows determining the shortest path of items or tools in order to change the schedule and introduce the organizational aids, the increase of productivity, the improvement in humanization of work, better utilization of the surface of a workplace: shortening the total path of movement of the object of work.

(3) *The method of stroke* studies the arrangement of equipment in the workplace when performing an operation. Graphical models of equipment are applied in the workplace and strokes of workers are drawn in the shapes of curves which represent the projection of body bearing on the plane of moving.

   A small circle is noted down at the beginning of a stroke, an arrow at the end. Strokes are separated by boundary points which represent the inaction or grips that do not belong to the grip of transport. The shortest total path allows the increase of productivity and humanization of work, better utilization of the surface of a workplace, reducing the total path of movement of workers which reduces the energy while carrying the object of work.

(4) *The method of a map of the sequence of grips on work object* studies the schedule of grips on the work object while performing an operation. The

actions undertaken are changing the sequence of grips on a work object while performing an operation; the elimination or reduction of certain grips (relief) are suggested while performing certain grips. The method of a map of the sequence of grips on work object allows the increase of productivity and humanization of work, especially the shortening of a total path of work object, reducing the number of grips, shortening the duration of the operation.

(5) *The method of spatial schedule and the sequence of grips* studies the sequence of grips while performing the operation in the workplace, as well as the arrangement of equipment. The change of the sequence of grips while performing the operation is carried out, the elimination of certain grips is suggested, together with actions to change the schedule. The method of spatial schedule and the sequence of grips helps to achieve the increase of productivity and humanization of work, as well as shortening the path of moving the work objects, shortening the duration, shortening the length of movement of workers.

(6) *The method of a map of grips* studies the sequence of grips while performing an operation as well as the arrangement of equipment and commands on the workplace. The change of the sequence of grips while performing an operation is carried out. The elimination or reduction of certain grips (relief) is suggested, together with actions to change the arrangement of equipment and commands. The method allows the increase of productivity and the improvement in humanization of work, better utilization of existing resources, shortening the total length of moving the work object, the length of path, the number of grips, the duration of the operation, the relative time of work.

(7) *The method of a map of movement* studies the interdependence of movements when performing an operation and the sequence of movements and arrangement of equipment and commands in the workplace. The change in the interdependence of movement is carried out; the elimination of certain movements is suggested, as well as shortening (relief) of their performance. The method allows the increase of productivity and humanization of work, better utilization of certain resources, shortening the total path of moving the objects of work, reducing the number of movements and shortening the duration of operations.

(8) *The method of a map of interdependence grips* studies the interdependence of grips performed during the operation in the workplace. The map of interdependent grips follows the process of work and not the object of work, therefore all the grips are grouped into three larger sets: work, transportation and waiting. The resources are shown on the abscissa, and the cumulative time on the ordinate. The change of interdependent grips in performing the operation is carried out, as well as the elimination

of certain periods of waiting, or their reduction and parallel work on various resources. The method allows the increase of productivity and humanization of work, proper utilization of existing resources and shortening the total duration of the operation cycle.

(9) *The method of movement* studies the arrangement of equipment and commands in the workplace, the sequence of movements and interdependence of movements of left and right hand. The change of the arrangement of equipment and commands is carried out; the change of the sequence of movements when performing an operation is suggested, as well as the elimination of certain movements and changes in the interdependence of movements of left and right hand when performing an operation. The method of movement allows the increase of productivity and humanization of work, better utilization of a working place, shortening the total length of the hand movements of workers, reducing the number of movements, synchronization of movements of left and right hand and shortening the duration of the operation.

4.5   Ergonomic design of workplace in garment industry

Different types of technology have the influence in different ways of working. For example, a line of clothing production is based on the measurement of partial working time, where a chronometer determines the average time for each work element within the time study and the motion study. That is how a man is brought into the rhythm of production line by a number of movements which can be well calculated, but certainly not the best ones for each individual worker.

Proper ergonomic design of each workplace, along with finding suitable methods of work with the appropriate time standards ensures better structure of technological operations with the increased efficiency of sewing machines. Working posture at sewing machines should allow the mobility of the limbs, ergonomically favourable arrangement of working and visible zones and a stable balanced state when performing the work process.

Technological processes of sewing clothes are performed on production lines with a large number of technological operations where each technological operation does not last long and has a significant psychological, physical workload for each worker. The material which goes through the process of work, due to its physical–mechanical characteristics, requires a careful handling when taking, assembling, positioning and putting it aside. Therefore, the structure of technological operations is mostly (65%) related to the handling of material within support-hand technological grips. The very processing on a machine (sewing grip) is performed during the machine or machine-hand time (25%), while 10% of time is used for non-production work. When designing workplace in the process of sewing it is necessary to achieve dimensional harmony of human–machine system of inter-phase transport, with the correct physiological posture of sitting, which allows rapid and accurate movements of the motor when switching on the machine and processing the work object, a high level of coordination of movements, a correct position of the spine and good position of the head. The posture of the body of workers, the complexity of the structure of individual movements within the performance of technological operation of sewing and the level of muscular and visual control of the worker depend on the type of technological operation, the type of sewing machine, its technical equipment, machinery and the layout system of workplaces.

When designing workplace in garment industry it is necessary to apply five principles of ergonomics:

(1)   Ergonomic principles in designing workplace.
(2)   Ergonomic principles in designing working processes.
(3)   Ergonomic principles in determining working time.
(4)   Ergonomic principles in handling material and tools.
(5)   Ergonomic principles in designing environment.

(1)   *Ergonomic principles in designing workplace*
   - Properly designing workplace should make possible for the work to be performed either in standing or sitting posture.
   - Workplace height should be such that a standing work can be replaced by a sitting one.
   - There should be enough space at the workplace for the operator to stretch his legs comfortably.
   - Each operator should have a seat of such type and height as to assume proper posture in work.
   - Armrest should be provided if the nature of work allows.

(2)   *Ergonomic principles in designing working processes*
   - In performing an operation, the posture which should be applied, requires the minimum energy consumption.
   - Work should be organised in easy and natural rhythm.
   - Standing posture should be used only when higher force should be applied by hands or when movements are necessary (cutting material, trim, Figure 4.2).
   - Work should be organised so as to use both hand simultaneously whenever possible.
   - Hand should be freed from work whenever possible and while serving the tools or machines done by feet (work on special sewing machines whenever possible, Figure 4.3).

4.2 Work on press.

4.3 Work on special sewing machines whenever possible.

4.4 Position of elbows at work.

- To make a movement maximally economical, it is necessary to employ adequate muscle masses.
- Sewing workplaces are shaped assuming that the worker has good visual skills, i.e. favourable working posture which consists of slightly bent upper part of the back with a work line of sight that can include flexible front of the head in a comfortable posture to a maximum of 30° and an additional eye rotation of 10°. This posture allows field of vision with the viewing angle ±1°, which achieves high visual acuity required for accurate management of technological operations of sewing.
- This is to specify the height of sitting, the height and the size of desktop machines, pedal position, distance of chairs, with the necessary sight and visual acuity and the ability to perform simultaneous movements of hands, legs and torso (Figure 4.4).
- The procedure of designing the workplace in a sewing room requires determining the angles of kinematic system, whereby a suitable position of the foot on the pedal of a sewing machine is at angle of 90–100°, while angles suitable for joints of under knee-upper knee are 90–110°, and upper knee-torso 90–95°. This is the way to get ergonomically functional and physiologically correct sitting working posture, with the proper arrangement of equipment and means of work, proper angles of vision and distances.

4.5   Wrong positions of work pieces.

(3)   *Ergonomic principles in determining working time*
  • Production time can be determined only for the operator who is skilled for the job and has average experience.
  • Pause for the handling loads, improper body postures in work and monotony should also be taught about when calculate manufacturing time, as they seriously impact fatigue coefficient.
  • Real coefficient and additional time, including lunch break, breaks for physiological needs and justifiable organisational losses should be calculated and included into the norm (see Chapter 3).

(4)   *Ergonomic principles in handling material and tools*
  – Operator should be free from transport procedures as much as possible.
  – Hand should be free from holding all the work pieces.
  – Each instance of handling the material should be provided it is economically feasible, mechanical or automated.
  – Tools, materials and work pieces to be handled should be positioned so that the operator is not required to bend his body, if possible (Figure 4.5).
  – Tools should be put at the workplace whenever possible (Figure 4.6).

4.6 Tools on workplace.

- Recommended force for a permanent lifting operation under favourable conditions is for men 176N and 98N for women. For occasional lifting under favourable conditions the force is 490N for men and 294 for women operators. If work includes permanent load bearing permissible load is 392N to 490N for men and 147 to 196N for women. This is important when transport and taking the textile material on cutting table.

(5) *Ergonomic principles in designing environment*
- When using both daylight and artificial illumination, the light source should always be to the left.
- Intensity, distribution and type of illumination should prevent excess strain of the eyes.
- It is necessary to find the appropriate intensity, timing and type of illumination that will ensure a smooth performing of the production process throughout the work day.
- Lights are placed on that way that 60% of light comes from the main source, while the other part comes from additional lights in each workplace. A number of other factors are also taken into consideration, such as the height of working space, the coefficients of reflection of surfaces of the workspace (e.g. table or wall), and the amount of

4.7 The light in a cutting room.

natural light in the room. The light in a cutting room is shown in Figure 4.7.

– Individual sources of light on sewing machine for work with dark materials and topstitch (Figure 4.8).
– Approximate illumination values are as follows: 60 lx rooms with low frequency; 120 lx light strain of the eyes; 500 lx normal strain of the eyes (for reading and writing); 1000 lx high strain of the eyes.
– Illumination values for cutting and sewing dark material 600 to 1000 lx.
– Illumination values for cutting and sewing lighter material to 500 lx.
– Illumination values for the CAD system ranges between 300 and 600 lx.
– Workroom temperature should be adapted to the type of work to be done, as normal functioning of the organism is closely connected with constant inner temperature of the body.
– Comfort is achieved at relative air humidity of 50 to 70 %, with the following temperatures in workrooms: 20 to 21 °C light, sitting office work; 18 °C light work in standing posture; 17 °C hard work; and 15°C very hard work.
– Air exchange in the room should be at least 45m^3/h per person, for light and 90m^3/h per person for hard physical labour.

4.8   Individual sources of light on sewing machine.

- Sewing room and finishing room require the usage of fans and air conditioning in summer. If they are used in warm periods, the difference between external and internal temperature should not exceed 7 °C and relative humidity from 40% to 60%. It is advisable that the temperature in a working day should vary in 1-2 °C.
- Minimal work area per person should be 4m². On Figure 4.9 there are dimensions of sewing machine.
- Noise and vibrations distract the worker from the work object causing tension and unrest, and if they last long they can cause fatigue and insomnia. Longer exposure to noise can lead to ear damage, whereas faster vibrations cause rapid heartbeat, increase blood pressure, reduce sight and make other problems in the body of worker.
- To organise work properly noise should not exceed 50 dB(A) for intellectual work, 70 dB(A) for office and similar work and 90 dB(A) for other types of work. Table 4.1 gives the allowable noise levels depending on the type of work.

Where: a – the noise made by a machine or device which is directly handled by the worker,

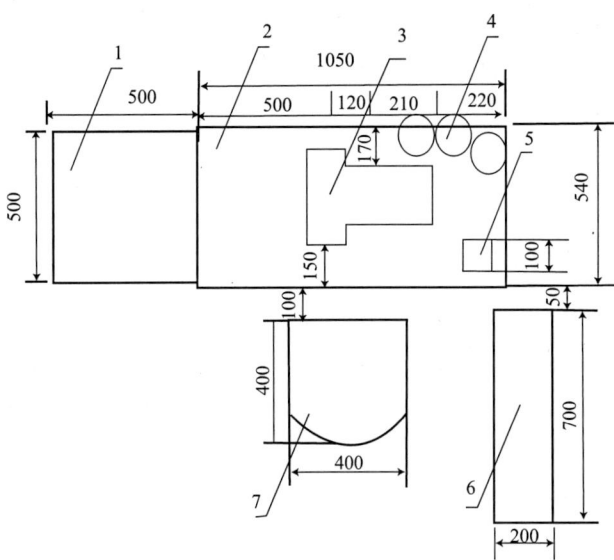

1. Slanted conveyor belt, 2. sewing machine table, 3. sewing machine head,
4. cotton reel stand, 5. label box,6. movable stand, 7. chair

4.9 Dimensions of sewing machine.

Table 4.1 The allowable noise levels depending on the type of work

Type of work	The allowable noise levels on workplace, dB(A)		
	a	b	c
physical work without requiring mental strain and perception of environment by hearing	90	84	80
physical work focused on accuracy and concentration; periodical monitoring and environmental control by hearing; driving of means of transport	80	74	70
work that is done by frequent voice commands and acoustic signals; work that requires constant monitoring of environment by hearing; routine work mainly of mental character	–	70	60
routine work mostly of mental character that requires concentration	70	64	55

(Continued)

mental work focused on the control of work of group of people who perform mostly physical work, work that requires concentration or direct speaking and telephone communication	–	60	50
mental work focused on the control of work of group of people who perform mostly mental work, work that requires concentration, right speaking and telephone communication; work exclusively related to talks over the means of communication	–	55	45
mental work that requires large concentration, exclusion from the environment, precise psychomobility or communication with a group of people	–	–	40
mental work related to great responsibility, communication to deal with a group of people	-	-	35

b – the noise made by a machine or device which is not handled by the worker,

c – the noise made by non-production sources (device for ventilation or air conditioning, other factories, street traffic, etc.).

– Colour in the working premises affects the feeling of warmth or coldness of the workers. Various tests showed that brighter colours have a pleasant effect on workers, increase their concentration, mood and speed of work, whereas cold and dark colours create a feeling of apathy, bad mood and sleepiness of workers.
– Music removes fatigue among the majority of workers, reduces monotony and anxiety at work, if the optimal duration of music is two and a half hours in one day, in intervals of 12 – 20 min.
– Proper hygienic conditions should be provided, as well as adequate number of rest room and adequate devices in them.
– Workplaces should be kept clean at all times.

References

1. Barnes R.M (1968), *Motion and Time Study: Design and Measurement of Work*, (7th ed) John Wiley, New York
2. Chapanis, A. (1996), *Human Factors in Engineering Design*. NY: Wiley, New York

3. Colovic G (2007), 'Oblikovanje radnog mesta kao faktor motivacije u odevnoj industriji', *Industrijski menadzment i razvoj,* FIM, Krusevac, 244–249
4. Colovic G and Petrovic V (2001), 'The analysis of working time losses in a technological process of the production of men's T- shirts', *1ˢᵗ International Ergonomics conference, Ergonomy 2001*, Zagreb, 143–153
5. Döring B. (1976), *Analitical Methods in Man – Machine System Development – in: Introduction to Human Engineering*, Kraiss Moraal,Verlag Tüv Rheinland Gmbh, Köln
6. Dul J and Weerdmeester B (2002), *Ergonomics for Beginners: A Quick Reference Guide*, Taylor & Francis, London
7. International Ergonomics Association (Available at http://www.iea.cc/ergonomics/ [Accessed 10 October 2009])
8. European Foundation for the Improvement of Living and Working Conditions (Available at http://www.eurofound.europa.eu/pubdocs/2006/98/en/2/ef0698en.pdf [Accessed 28 September 2009]).
9. Murrell K. F. H. (1965.), *Man in his Working Environment,* Ergonomics, Chapman & Hall, London
10. Stramler J.H (1993), *The Dictionary for Human Factors/Ergonomics*, Boca Raton, LA: CRC Press
11. Wilson J (2003), *New Methods in Ergonomics*, New York
12. 90/270/EEC – Council Directive on the minimum safety and health requirements for work with display screen equipment

5

Analyze of the planning, layout and logistics in garment manufacturing

Abstract: The industrial mode of production of clothing requires thorough preparation of the production process, because several factors are to be connected: people, time, machines, production facilities, organization and material in a coordinated and rational system. The technological system of production of clothing must provide the required product quality, the production volume needed, the delivery of finished garments within the stipulated time and the maximum utilization of capacities with minimal costs.

Keywords: computers, production preparation, planning, layout, logistics.

5.1 Analyze of the planning, layout and logistics in garment manufacturing

The dynamics of international strategy of the fashion industry is speeding up with the development of new business concepts and requirements of the increasing changes in the market. The world itself is viewed as a potential source of production, being at the same time a unique market of clothes. The base of the economic system lies in the fact that the ultimate success is positive with those manufacturers whose products are cheaper and better than the competing ones. Higher productivity allows the increasing taking part in the global market. Those who can not keep pace lose their place. Final customers do not care where the product comes from, but the parameters which decide about its choice (quality and price, etc.). Nowadays, the clothes is purchased as a daily item with a short period of usage.

Production planning is a key phase of the management process and the overall development of the organization and the safety in performing tasks depend on the quality of plan. The fact is that planning saves time and enables various resources of PBS to be used in the best possible way. Planning can be defined as setting goals, and refining activities, routes and resources for their realization. After a set- up and refinement of goals, the planning process should be systematic but also flexible. Plans may be strategic and operational.

Management of PBS has a unique responsibility – one of them is making strategic (long-term, corporate) plans which, depending on the activities of the organization, are made for a period of one to five years and more. Strategic plans refer to long-term objectives of PBS and they include:

- Existence of mission and vision (the definition of work that deals with PBS).
- Definition of corporate goals (accurately determined increase in absolute amounts, percentages, etc.).
- Assessment of the external environment (economy, competition policy, market trends, etc.).
- Assessment of internal resources (analysis of strengths and weaknesses of PBS).
- Analysis of possible sequence of events.
- Turning the plan into concrete activities (refinement: priorities, the framework of deadlines, resources, responsibility for implementation and coordination of programs, etc.).

Operating (management) plans start where strategic plans end, aiming to realize them through practical and more elaborate activities and measures, and they are related to shorter periods of time (daily, weekly, monthly, quarterly and semi-annual). These plans function as strategic plans and they are all about performing the scheduled work on time without using more resources than anticipated, in order to avoid crisis situations and additional costs they carry.

Operating plans can contain other plans:

- investment plan,
- plan of staff,
- plan for education and innovation skills,
- marketing plan,
- procurement plan,
- sales plan,
- production plan,
- holiday plan, etc.

The task of preparing the production is to determine all the circumstances of production so the process could carry out normally, with no improvisation. Preparation of production includes: constructional, technological and operational preparation. Technological preparation includes determining the technological process, the selection of machines, tools and determination of material quality. Operational preparations include: developing plans for individual production facilities, determining the amount of material, determining the time of manufacture, production of necessary documents and so on.

Besides the technological preparation, operational preparation has very important tasks in the overall construction output in garment industry. The operative preparation usually consists of the following tasks: determining the capacity of production, production planning, production monitoring, planning the necessary quantities of materials, launching work orders, scheduling of production and assembly of data for plan calculation.

Production capacity is the amount of product that a manufacturing section can implement for a certain period of time. The production capacity is affected by many parameters of which the next are most important:

- type and amount of funds that are available in a production facility,
- number of workers and their qualification structure,
- type and quantity of items of clothing i.e. products that are produced and
- market.

If PBS has modern and highly productive means it can count on a large quantity of products and services that can be achieved for a certain period of time. Accordingly, means of work are decisive factors in the restriction of production capacity. In addition to the funds, a number of workers has a very important role as well as their qualification structure for determining the capacity of production. If there is not a sufficient number of workers, then the available funds will not be used sufficiently and planned tasks can not be achieved. Likewise, if workers are hired more than the required number, under-utilization of workers occurs. Thus, the human factor greatly affects the determining of production capacity.

The production capacity is also influenced by the objects of work, considering not only the types of garments that are produced, but also the possibilities of getting the necessary amounts of material. Every delay in production due to the lack of materials causes non-using of capacities, delivery failure, increasing production costs and worker dissatisfaction.

The market has a very important role in determining the capacity of production. If the market has no need for specific products, then it is a limiting factor of production capacity. Only the attractive products can be demanded on the market and that is why PBS should be orientated towards producing such products.

Annual production capacity is determined by the formula:

$$C_g = \frac{D \cdot T_d \cdot R}{t_1} \qquad [5.1]$$

Where: D – number of working days per year,

T_d – daily fund of working hours,

R- number of employees and

t_1- time production per unit of product.

Daily production capacity (C_d) is calculated according to the formula:

$$C_d = \frac{T_d \cdot R}{t_1} \qquad [5.2]$$

or

$$C_d = \frac{C_g}{D} \qquad [5.3]$$

If different products are planned to be made, then partial annual (C_{gi}) and partial daily capacities (C_{di}) are calculated according to formulas:

$$C_{gi} = \frac{D \cdot T_d \cdot R \cdot U_i}{t_1} \cdot 100 \qquad [5.4]$$

Where: U_i – the participation of certain products as a percentage of the total quantity

$$C_{di} = \frac{T_d \cdot R \cdot U_i}{t_1} \cdot 100 \qquad [5.5]$$

or

$$C_{di} = \frac{C_{gi}}{D} \qquad [5.6]$$

The sum of all partial capacities gives the total annual, i.e. daily capacity:

$$C_g = C_{g1} + C_{g2} + C_{g3} + \ldots + C_{gn}$$

$$C_d = C_{d1} + C_{d2} + C_{d3} + \ldots + C_{dn} \qquad [5.7]$$

However, capacity can be given in temporal indicators, or through the annual fund of working time and daily fund of working time.

The annual fund of working time can be calculated by the formula:

$$T_g = D \cdot T_d \cdot R \qquad [5.8]$$

The daily fund of working time is calculated according to the formula:

$$T_d = T_d \cdot R \qquad [5.9]$$

or

$$T_d = + \frac{T_g}{D} \qquad [5.10]$$

Where: T_g – annual fund of working time,

T_d – daily fund of working hours.

If the data on a partial number of workers engaged by the items are well-known, then the partial annual fund of working time (T_{gi}) is calculated by formula:

$$Tgi = D \cdot Td \cdot Ri \qquad [5.11]$$

There are usually three types of capacities in garment industry, and they are: borderline (installed, built-in), planned and actual.

Borderline capacity (C_g) includes the potentials of some production unit to produce a certain quantity of clothing products in a certain period of time without paying attention to any losses.

Planned capacity (C_p) is a capacity that is determined by taking into account the losses of machines and workers (losses at overhaul, repair, etc.). Thus, for example, when planning the losses in the work of machines eight percents can be taken and the coefficient of losses is 0.92, whereas talking about workers it is four percents, i.e. the loss coefficient is 0.96.

Actual capacity is the achieved quantity of product for the planned period of time. If the actual capacity (C_s) is within the limits of borderline and planned capacities, the production is expected to take place normally. However, if the actual capacity is less than planned, some organizational steps should be taken in order to increase production.

In case when the actual capacity is larger than the borderline one, then the latter is cither set incorrectly, or there was a violation in the quality of production. If there was a violation in the quality of production, some organizational steps must be taken in order to ensure the planned quality of apparel products.

When the volume of production which needs to be achieved in a certain period of time is determined, then such a volume burdens available capacities so these things may occur:

– bottlenecks in production and
– free capacities.

It often happens that the customer service agrees upon larger quantity of products and that a working unit in some workplace is not capable of realizing a certain quantity of products of required quality.

5.2   Application of computers in preparing for the production of clothing

The process of forming a garment in garment industry requires comprehensive technical and technological preparation of production. The greatest number of errors, and thus the costs associated with product quality, arise in defining a garment, developing a product and planning of technological process of making clothes. 75% of all errors that appear on the product are believed to occur during constructional preparation. Costs of avoiding errors are the lowest ones when defining and developing a new product – a garment or a collection of garments, especially if you use the CAD (Computer Aided Design) system.

Constructional preparation includes a number of jobs which largely influence the final product, its functionality and the price of product.

The jobs of a designer and a modeller in constructional preparation are, mostly, highly professional and demanding, because the quality of clothing includes, among aesthetic and functional requirements, the stability of shape, simple maintenance, feeling nice while wearing, the beauty of fall, etc. Physical and mechanical characteristics of textile materials which are used to make clothes have a big influence on all that, as well as on the construction of clothes. Giving forms must be systematical and continuous, which means that within the framework of industrial production, there must be a program of work that includes:

- The basic definition of the problem of design.
- Specification of a problem.
- The concept of designing within all current features and specific factors, such as:

(a)   function of a garment,
(b)   body measures, dimensions of garment, standards and tolerances,
(c)   structure and functional complexity of garment and
(d)   structure of materials and production technology.

Construction preparation for industrial production covers activities which can be divided into several groups: collecting ideas for new products, designing and sketching models, a description of the model, the basic construction, standardization of new cutting parts, modelling the basic structure on given creation, completion cutting parts, grading of cutting parts, cutting the pattern for tagging, making cutting layout and prototype of production models.

The examples of developing systems which are present on the world market (Gerber, Lectra, Investronika, Optitex, Assyst) allow taking all variations of future garment into consideration. Their behaviour is simulated, their impact on the environment is analyzed, simultaneous design is used aiming to obtain the best solution in the given conditions which reduce subsequent changes (correction model design, adjustments to fit in cutting layout for basic and supplementary material) to a minimum. Creating a prototype model for industrial production in the formation of products in the clothing industry involves the development of CAD methods for the preparation of construction segments. The number of possible variations in

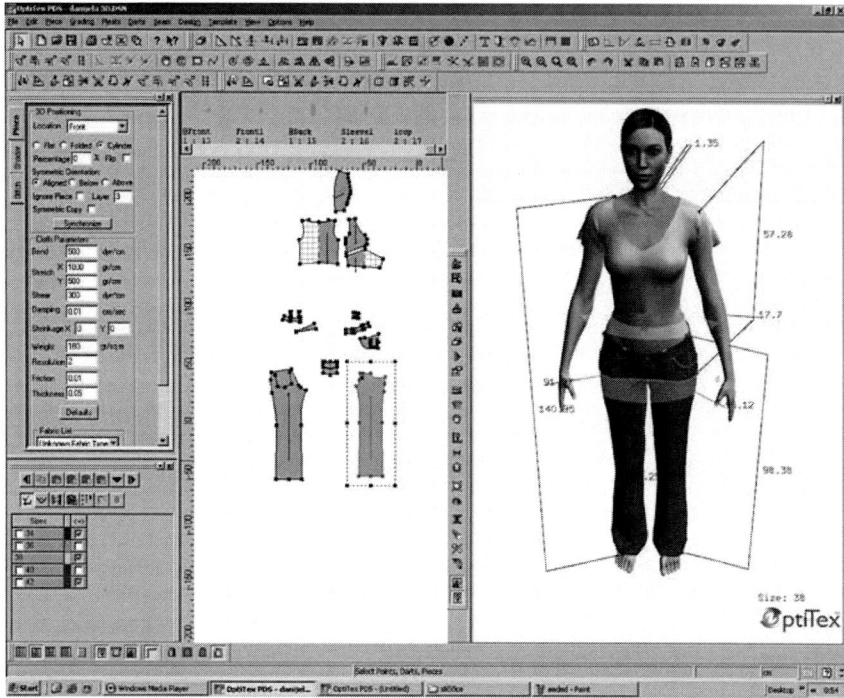

5.1 Control and positioning of 3D objects and adjusting cutting parts with parameters of prototype.

colour, shape and design, modelling and re-modelling of garment is almost unlimited (Paunovic, 2009), Figure 5.1.

Computer simulated clothing is a truly designed textile material redesigned from two-dimensional form into three-dimensional one. The progress of research about the behaviour of textile surface products affected by various forces while wearing provided a computer simulation of fall and fabric crimping, so called draping and visualization of finished garments. Knowing the type of material a garment is made of and its construction can lead to a fairly realistic picture of a garment and its graphic presentation on a virtual model of man. By simulating the fall of the fabric and its dynamic changes some possible deficiencies before cutting and sewing can be noticed and corrected, Figure 5.2.

CAD systems in garment industry help to improve development, analysis, redesigning models and making decisions about solutions for clothing products. True solutions are optimal solutions which satisfy the requirement of getting maximum quality for minimum price. Designing products is a critical activity of the manufacturing process because it is estimated that its share is 70% to 80% of the cost of development and production.

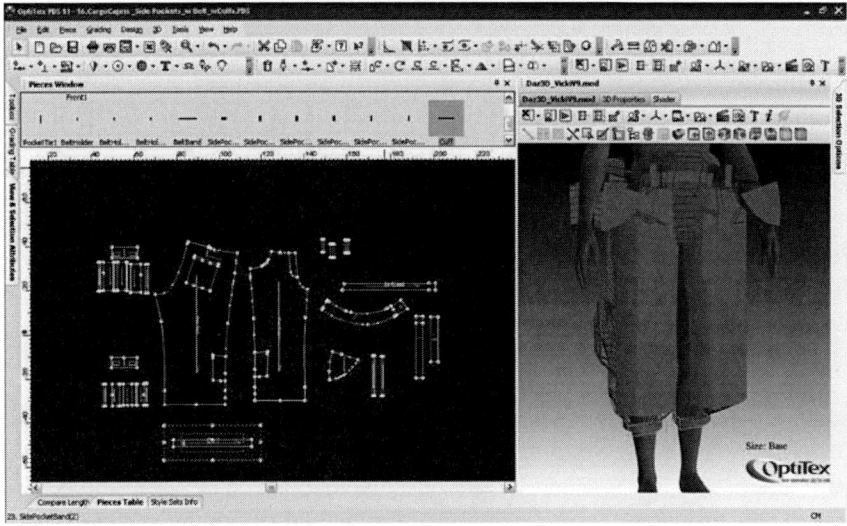

5.2 Pattern making and virtual sewing.

Geometric design cloth pattern is a complex procedure with many reps and interdisciplinary programs whose priority is reflected in:

– flexibility of dimension,
– increasing accuracy,
– increasing productivity,
– possibilities of real visualization,
– optimization of construction,
– optimization of deadlines of production preparation,
– security and quality management,
– planning, monitoring and efficiency in creating products.

In computer technology, there is the notion of intelligence as a totally new attribute of a manufacturing process. Intelligent manufacturing processes are required:

– the ability of fast, cheap and productive adaptation to new requirements and situations,
– the ability to learn and use one's own and other experiences quickly,
– the ability to reason and understand causal relationships and
– disclosure and registration of production parameters and their incorporating into adapted behaviour.

Simulation process of construction preparation allows the virtual product development by: the exchange of electronic data on material, model, size and personalization of virtual model of products and customers, quality control, data transfer to CAD systems, the possibility of realization of CIM (Computer Integrated Manufacturing) concept and flexible methods of production and optimization process. Computer support of small serial production by simulation on the CAD computer system provides answers to the dynamics of changes of fashion trends, optimization of product development and process and direct involvement of customers into the process of creating the final product.

Designing fashion products using CAD systems improves the development, analysis, redesigning models and making decision about true solutions. True solutions should be the optimal solutions that satisfy the requirement to get maximum quality for minimum price.

Cutting layout is a set of cutting parts of one or more types of garments, rationally put in the rectangular surface of material with endings in a right angle. Its purpose is to determine the path which will contain layers of materials and to reduce their consumption. Garment quality requires its width and direction of the base and weft. Cutting layout can be one-size and more-size. Their production also depends on the method of putting cutting layers (spreaded or folded material...), as well as on the type of material used. CAD software packages for fitting cutting parts into cutting layout were developed in order to increase productivity by reducing labour and material investment. Using software in development cutting layout with manual, automatic and interactive creation of cutting layouts can lead to optimization with large speed and reduced consumption of materials, achieve maximum flexibility while fitting cuts, make changes in quantity, size, and number of layers, view all the parameters at any time of the process of making. In addition to that, it is possible to obtain results and save time with calculations obtained without fitting cutting layout, in order to optimize the cutting plan (Figure 5.3).

Well-known manufacturers of CAD equipment for construction preparation have produced automatic machines for putting cutting layers, which are directly connected to the CAD work units and the aggregates for cutting.

Apparel industry of developed and medium developed countries in the world is going through many crises and steady decline in production, loss of markets, dismissal of workers, closing factories and moving production to underdeveloped countries with cheap labour. Developed countries were forced to intensify the research of automated system of technological processes, which led to progress in the area of mechatronics, automation and robotics, and the invention of new so-called intelligent sewing machine. Modern sewing machines have programs and programming interface to connect to a computer, which is the reason for setting blocks adjusted to the machine for two-way communication with computer. So the instructions from computer are sent directly to the worker or the machine in case of correction, the sequence of activities and others. In the opposite direction, towards the computer, data about the state of machinery, progress of work, etc. are being transferred. Data collected from the machines

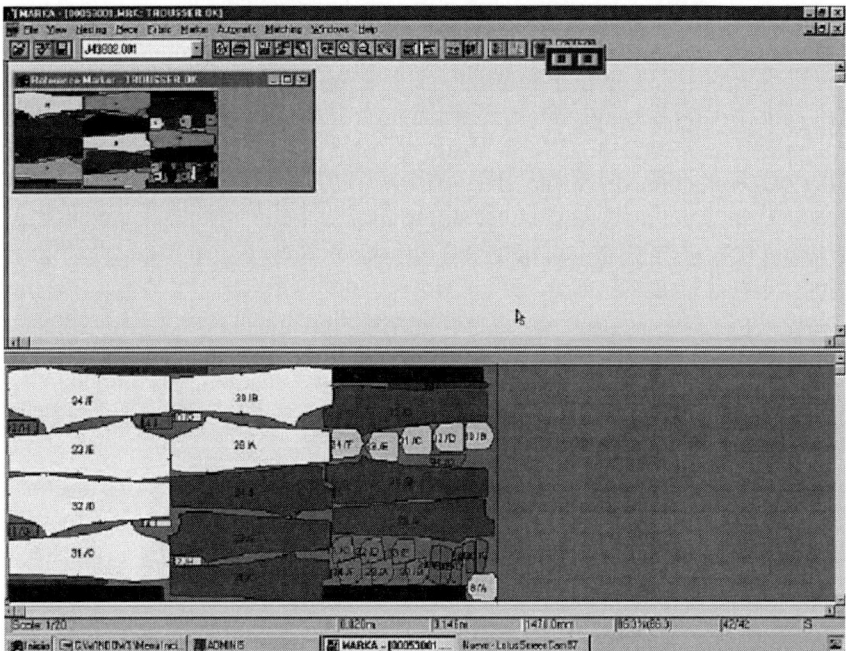

5.3 Making cutting layout (marker).

in the network are analyzed by the computer with the help of special programs and puts them in a database, where the data can be used by various users with specific application programs.

Modern fashion design requires a small amount of clothing, many colours and patterns, so the production plants daily deal with many work orders, which caused making of technical documentation to be one of the biggest problems in clothing industry. Computer for the organization of cutting technological process can be connected to a computer for making technological documentation and work order, and through CAP (Computer Aided Planning) of system the planning of clothes is carried out as well as sending information about the material on the basis of which sewing machines are programmed (material transport system, regulation thread tension, the force of pressure foot, sewing speed, etc.) intermediate transport and data about cutting parts for conducting of CNC (Computerized Numerical Control) of sewing machines.

The application of computers allows, in trimming phase, achieving the quality of ironing by selecting the optimal parameters of ironing and the saving of energy (electricity, water, steam, air). Programmed finishing machine allows maintenance and measurement of pressure, temperature, vacuum air and compressed air for ironing.

CIM includes computer integration of business, engineering, manufacturing and management information that links all the functions of the company from marketing to distribution of products. Benefits of CIM system are: fast response to market demands, easy modification of fashion products, reducing production cycle, high quality products, low price of production, increasing flexibility of production, rational utilization of textile materials, production equipment and workers. CIM is a concept of production in which the whole flow of production, from the entry of raw materials to the exit of finished products into the market, connected, monitored and controlled by computer systems that are in conjunction with all the processes that take place in a production business system. Data which are collected and processed are: the entrance of new materials in the warehouse, the situation in the warehouse, the cost of production, data on workers, selling products, the situation in production and others. Traditional organizational structure is not adequate for the concept of CIM, so along with the development strategy of CIM the profile in the clothing industry must be changed.

CIM concept has been introduced in the garment industry for the last ten years and up to these days it has been fully integrating the construction preparation and cutting technological process of clothes. In Figure 5.4

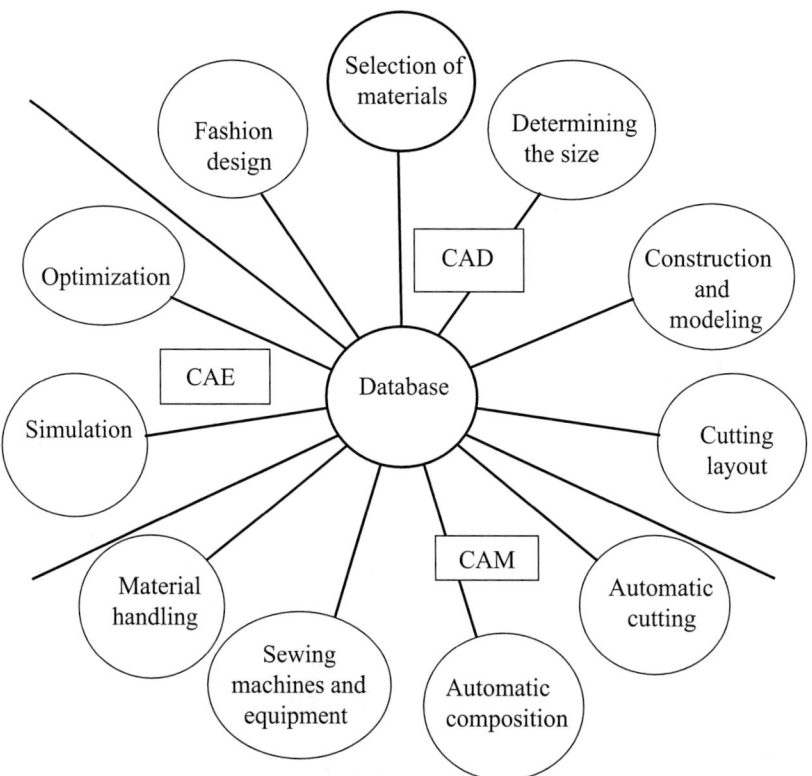

5.4 Interactions CAD, CAM and CAE in the garment manufacturing.

interactions of CAD, CAM and CAE (Computer Aided Engineering) in the garment manufacturing are shown.

Formation of integrated CAD/CAM (Computer Aided Design/Computer Aided Manufacturing) in a CIM concept achieves the increased quality of production for two to five times, the reliability of manufacturing operations from 40% to 70%, reducing costs for 20% and increasing the ability of engineers in the preparation and the analysis of process for three to 30 times.

The introduction of CIM concept leads to connecting to a common manufacturing system with common goals in a PBS:

– reduction of production costs,
– increasing flexibility of production and
– increasing quality.

The advantages of computer technology in garment industry, especially those whose production is within the CIM concept, can be presented at several levels:

– Improving working conditions in garment industry is achieved through collection, processing and automatic evaluation of data, reducing the "manual" work, better organization of labour schedule, better collection and analysis of information about customers and suppliers, easier communication between the structures of PBS, a faster and more efficient data processing, simplification of various operations, the availability of data from any point in the world with internet, as well as optimization and systematic flow of information.
– Better utilization of production capacity is achieved due to better surveillance over machines, schedule of labour, determining conditions of optimal task settings, reduction of waiting time because of better planning and faster obtaining more accurate data on the total quantity of production which is, therefore, cheaper.
– Improving productivity stems from the interactive relationship between workrooms and services of planning or making orders. Possible delays in production are currently registered, so one can respond in time and examine better ways of production in order to improve it continuously. Production can be planned in its very progress.
– Improving the quality of clothing is due to rapid perception and elimination of errors during production, and therefore reduces the subsequent quality controls and losses of time.

Modern industrial garment production requires continuous innovation and improvement of production processes, technology and work quality. By introducing information technology, using knowledge of expert systems, artificial intelligence and simulation of partial or complete technological process, planning and monitoring of production, new ways of designing and techniques of garment making have appeared. Although computer-integrated manufacturing of clothing which includes CAD, CAM and CAPP (Computer Aided Process Planning) connects all vital functions of technical and production systems, from storage of raw materials through the technical preparations to the finished goods warehouse, it is necessary to create a preparation for:

- unpredictable market movements,
- expansion of nanotechnology,
- construction of new textile materials used in all aspects of life (medical textile, geo-textile, ambient textiles...),
- increase the frequency of introducing new products,
- changing of parts of existing products,
- large and frequent changes in fashion trends in apparel products and their combinations,
- changes in the application of regulations (standards, safety, environmental, application of eco-textile...) and
- changes in the technological process itself.

One of the main reasons and objectives of introducing the concept of CIM is the expectation that equipment and facilities should be used up. The ultimate goal of CIM is a total informational and technical control and integration of logistic activities in the industrial system from the time of order to the delivery of finished products.

5.3   Risk Analysis

In order to understand these terms: risk management and risk management system, it is necessary first to define the concepts related to the risks. Risk is defined as calculation forecast emergence of negative events (hazards) that cause loss or calculation forecast emergence of positive events (opportunities/ chances), which bring us benefits.

Risk is a condition in which there is a possibility of negative deviations from the desired outcomes that we expect or hope will happen. In terms of business, risk is an unfulfillment of desired business objectives and it must include threats and opportunities from the environment that can potentially contribute to the growth and development of the PBS, but prevent development, and thus endanger the very survival of a PBS. The risk, in the broadest sense, is a particular danger, uncertainty, loss, or the uncertain future event that may have unintended consequences. The concept of risk comprises three elements:

- The perception that something might happen.
- Probability that something happens.
- The consequences of what might happen.

Risk consequences on the planned garment production are:

- exceeding the framework of the assessment of production costs,
- exceeding the requested date of making
- not acquiring the quality of clothing.

Tracking disorders and disturbances in production, their systematization and statistical methods can lead to the sizes which belong to risk and can be included in the calculation.

The risk may appear:

- because of placement of low bids in the market and
- in production compared to the size of the series that is produced.

Activities for achieving goals in the process of preparation and production:

- knowledge of market and adapting to conditions ,
- finding real costs,
- designing the organization of production for each garment,
- undertaking organizational and technological measures that will enable the production according to the planned costs,
- continuous control of production costs and
- creating a base of information for management of a PBS on the basis of production control.

Modern design at the beginning of the 21st century longs for refined lines of geometrical forms and aerodynamic surfaces burdened by the demand of assembly-line and mass production. This is not the case with fashion industry that dictates a great number of models and different styles with indefinite construction parameters of textile surfaces the garments are made of. On the other hand, there are producers who want to increase productivity and standardization of products and at the same time satisfy the consumers' needs as far as quality and diversity of products are concerned. The fashion appears in short intervals, it lasts for some time and disappears. It is repeated in cycles, slightly changed and modified. Many countries in their fashion centres with strong designer and marketing support allow creating of fashion whims that last for a short period of time on the markets with a large offer, competition and with consumers who are able to buy.

The consumers who are under the influence of fashion whims are impulsive category of buyers. The other type of consumers cannot follow fashion whims, but in a certain period of time that comes after a fashion whim there remains a fashion trend which has a long-lasting effect and allows to be worn by consumers with lower purchasing power. Lower quality products, if in trend as far as form, colour or ornamentation are concerned, are accepted by the consumers with lower purchasing power.

Buyers of garments want to be different, but not too much, so they accept basic fashion trend determiners with constant improvisation and expression of personal image.

Garment manufacturers in fashion industry are divided into leaders and those who copy (copyists). Leaders design a new product, impose and dictate new fashion helped by powerful centres of design and modern technology together with marketing company. Manufacturers who copy are late with production, but find their place in satisfying a large garment production market. They don't often strike back with the quality of products but there is a competition between them. Consumers react to fashion depending on their purchasing power. Fashion trends have an economic, psychological and social influence upon consumers. Consumers will often buy a new and prestige product that is above their purchasing power, particularly special brand of garment, wanting to become equal to a special social group by wearing such clothes and trying to achieve special social status.

Designing and getting new ideas in fashion industry today should become an organised process that requires thorough, systematic approach that is given a basic direction both by a designer and a capable manager team. Such an educated and creative team ensures that designing and gaining new articles of clothing, as being the most important element in the development of garment industry, should be done as planned and continually according to current situations and priorities. If this process is neglected, so many firms will struggle for their existence.

It is necessary to do SWOT (Strengths, Weaknesses, Opportunities and Threats) analysis in order to get marketing weapon and overcome all existing problems, which means to take all advantages and eliminate weaknesses of the PBS. It is possible to point out solutions and consequences, as well as to show how a fashion industry should get better with the help of this analysis. SWOT analysis must be summarized, specific and to analyse key questions of each PBS but not ad-hoc suppositions that do not point out future actions. SWOT analysis is the method of strategic planning that enables analysis of estimates and combining of internal factors with information from external sources on the market and in business environment.

Internal estimate as well as analysis of strengths and weaknesses in garment manufacturer refers to factors that can be controlled within the organisation. That concerns not only material expenses and technical- technological equipment but the reputation of the firm and existing brand, innovative activities of designers' team and marketing service. While doing an internal estimate there are hindered factors such as subjectivity, lack of trust, running away from reality. Weaknesses cannot often be overcome in a short period of time.

External estimate implies those possibilities and dangers that will have a main influence on business results of a textile firm. New fashion trends, i.e. new designs that should be accepted, should also bring new results. The problem which exists in garment industry is that we analyse and follow trends that have already taken place on the fashion scene so while a collection is being accepted and the preparation for production is getting completed, a new fashion demand is here, and the old one hasn't been accustomed to yet.

External estimate deals with advantages and dangers connected with a market, technology, scientific-technological development, changes in micro and macro environment, economy, ecology…

There comes the case when one textile firm, depending on its ability to follow fashion trend "blindly", sees the production of the very up- to- date design as advantage, whereas the other textile firm that has e.g. classical brand in clothing sees the same fashion trend and a new design as danger.

Possibilities of performing internal and external estimate in SWOT analysis are shown on the Table 5.1.

Table 5.1 Internal estimate (strengths and weaknesses) and external estimate (possibilities and dangers)

Internal estimate

Factors subjected to estimate	Characteristics of analysed factors
People	Skills, training, attitude
Organisation	Structure, relations
Systems	Formal/ informal, hand, computer
Communications	Characteristics
Products	Expenses, quality, life cycle
Production	Capacity, qualities
Finance	Profitability, liquidity, operating capital, indebtedness, solvency
People	Technical, commercial

External estimate

Market	Growth, fall, market share, fashion trends, target market
Technology	Development of products, distribution, manufacturing technology
Economy	Export/import, state of local currency
Society	Employment, trade-union practice
Legal regulation	Consumers' protection, pollution
Ecology	Energy, raw materials, recycling, environmental protection

Within a fashion PBS, paying special attention to the problems of designing a new product, it is necessary to observe within internal estimate first of all skills, training, attitude of marketing team, fashion designer and technologist and then the production management structure of the PBS itself, current systems (CAD/CAM systems), relations and communications among people. In order to manufacture an adequate product real expenses must be estimated, especially textile fabrics and quality of manufacturing garments, in order to lengthen a life cycle of a fashion product and not to neglect technical-technological capacities of the firm. The rhythm of technological development is faster than the development of human generation (life time for technological generation is four to five years, which is seven to eight times less than an engineer's length of service in clothing industry), so permanent training of production-technical employees is necessary.

Within an external estimate, when analysing possibilities and dangers of a fashion product that should be designed and redesigned in accordance with a fashion trend, it is necessary to study the very situation on the market, to determine target groups and categories of consumers. Garment manufacturers who do not invest in the development of products and production technology get into danger and can't "keep pace" with fashion trend although they try hard. A good distribution net, besides placing goods on the home market, should think of doing it on the foreign market. Because of poor economic power and large competition on the world market there are possibilities for many fashion PBS to fail, therefore a good distribution net should enable joint breakthrough for many fashion PBS to the foreign market, as the only real possibility.

A large number of not standardized fashion products, not harmonized sizes and quality of production are a danger for gaining access to the world market. SWOT analysis can be of a special importance in fashion industry when used for designing a new article of clothing, i.e. collection. The suggestions for the use of SWOT analysis when designing a new article of clothing in fashion industry are given in the Figure 5.5.

During the manufacturing of basic SWOT analysis different problems may appear and they can be solved (Kotler, 2006) with more critical POWER SWOT tool:

P = Personal experience

It provides the SWOT analysis to be based on the experience, knowledge, skills, attitudes and beliefs, because all the observations and personal feelings have an impact on the SWOT.

O = Order

Marketing managers often confuse features and advantages or disadvantages and risks, because the line between them is sometimes invisible.

W = Weighting

STRENGTHS	WEAKNESSES
- Futuristic design - Good image - Reaction to a new fashion trend - Quality of textile fabric and production - New ergonomic form of model - Short period of development of a model and short period of duration - Automatization of production processes - Industrial training conducted by specialist - Ecological requirements	- Very high price because of fast changes - Small series with a large number of models (three to five articles in work order) - Manufacturing of only three sizes - Bad covering of foreign market - High price of energy - Condition and price rise of raw material because of introducing VAT - Short time for optimalization of products
OPPORTUNITIES	THREATS
- Consumers' wish for new designs - Marketing of products into a new market - Market - Establishing "show room" objects - Making e-mail catalogue - Value of labour - Production of garments Made-to-Measure	- Import of similar articles of clothing at low prices - Competitors have lower price - Competitors have better distribution net with more sales places - Quick obsolesce of technology

5.5 SWOT analysis for new collection.

Certain elements of the SWOT analysis are often difficult to estimate, because some points are slower than other ones. Therefore, it is necessary to use percentage participation, for example: 1 = weakness of 20%, weakness 2 = 70% and weakness 3 = 10% (total 100%).

E = Emphasize detail

Many details, reasons and justification have been omitted in the SWOT analysis. For example, the technology itself is often both the weakness and the advantage.

R = Rank and prioritize

All the facts presented in the SWOT analysis have the impact on the strategy and that is why it is necessary to classify (rank) them from the highest to the lowest impact and thus determine the priorities. For example, if possibilities of a manufacturer for a new collection of clothing:

- The desire of customers for new creations = 25%
- Creating a "show room" facilities = 60%
- Launching new products to market = 20%

Then the marketing plan must be identified according to the following priorities (the opening of "show room", new creations, placing new products on the market, etc.). For ranking and setting priorities a Gap Analysis can be used, a simple and effective analysis that enables marketing managers to decide upon the appropriate strategy and tactics.

5.4   Optimization of planning

Planning means predicting tasks in a future period of time. Planning shouldn't be done as desired, but according to the real possibilities, including all possible potentials.

Planning is a complex job and it can be overcome only by good organization of work and following the specific methodology. In the PBS, there are:

- individual plans and
- overall plans.

Total plan consists of individual plans, such as:

- investment plan,
- production plan,
- plan of procurement of material,
- plan of required number of workers,
- plan of realization of products,
- plan of depreciation, etc.

According to the time period, there are:

- long-term plans, (including a period of five to ten years),
- medium term plans (including a period of one to five years) and
- short-term plans (including a period of one year).

Short-term production plans are most represented in garment industry due to the fashion trends that are often changing. In the short-term plans there are:

- seasonal plans or plans which include a period of six months,
- monthly plans which include a period of one month,
- weekly plans and
- daily plans.

Regardless of the time period, the production plan should include:

- name and type of products,
- the unit,
- the planned amount in pieces and financial indicators,
- normative material consumption,
- normative consumption time,
- overview of capacity,
- measures for carrying out the plan of production, etc.

Production plans depend on the length of production cycle. Production cycles in garment industry include the duration of the production process through stages for cutting, sewing and finishing. Therefore, the production cycle means the duration of the technological process of making garment from the first to the last operation.

The duration of the production cycle depends on several factors: selected system of technological processes, a way of keeping the process, the size of bundles, etc.

Depending on the duration, the cycle of production can be:

- The longest (occurs when the next stage does not begin until the previous is completed).
- Shortened production cycle (occurs when the part of the process of cutting finishes and the process of sewing begins, then parallel process of cutting and sewing until the cutting process is completed, and after that the sewing process continues until its end and then the process of finishing starts).
- The shortest cycle of production (occurs when the part of the process of cutting ends and the process of sewing begins, and when the part of the process of sewing ends the process of finishing starts, so there is a simultaneous conducting of three phases in a certain period of time).

Concerning the fact that it often happens that one production unit makes many different models of clothes during one month, you need to know the time for making every particular model, then the average level of productivity and the number of available workers.

The daily production of clothing deals with many work orders, so it is necessary to complete Scheduling (shop floor control). Scheduling means determining the order of carrying out jobs. Priorities can be determined on the basis of several rules:

- FIFO (first-in, first-out),
- LIFO (last-in, first-out),
- DDATE (earliest due date) – which product must be made first,

- CUSTPR (highest customer priority) – first the product for priority customers,
- SETUP (similar setup) – first the similar products which require a minimum setting of machines,
- SPT (shortest processing time) – priority for products that last for the shortest period,
- LPT (Longest Processing Time) – priority for products that last for the longest period.

Priority DDATE can have variations:

- SLACK (Slack minimum) – priority for jobs that have less time margin:
- CR (smallest critical ratio) – priority for jobs with smaller ratio of the remaining time to maturity and remaining processing time:

If CR > 1, then the product is made before the deadline
If CR < 1, then it's late
If CR = 1, then it is done on time

Already deployed operations often change their order, since there are new jobs coming into production. Scheduling is performed according to the above rules, which can be modified in a way that corresponds to the number of resources or the complex global rules:

- Expanded SPT (shortest processing time) – jobs are divided into A, B and C, according to their duration. Jobs A are performed with a minimum duration, but every few hours the production is interrupted and job B is performed. Jobs C are performed every day each.
- WINQ (work-in-next-queue) – sees the duration of performing job at the next resource.
- NOPN (fewest number of remaining operation) – according to the number of resources that are yet to be used.
- S/OPN (Slack per remaining operation) -according to the time reserve for the rest of the job.
- RWK (Remaining work) – a variant of SPT, the duration of performing whole job on all resources.

Resources which are bottlenecks have a negative effect on the process efficiency, because they limit its real capacity and do not provide a high-quality, fast and flexible production. Goldratt's limiting theory (Goldratt E.M 1997), defines a bottleneck in the process or in the BPS, because it assumes that the goal of every PBS is to "create money". Limiting theory is what prevents the system or process to reach a higher level of performance, i.e. it focuses on real capacity, inventory and production costs. Goldratt's theory coordinates production flow with demand according to the following principles:

(1)   it is necessary to harmonize the flow of the process, not the capacities of phases of the process,

(2) efficiency of bottleneck is not determined by its capacity, but by other constraints,

(3) utilization and use of resources are different concepts,

(4) an hour wasted due to the bottleneck is an hour lost for the entire process,

(5) an hour saved at bottleneck is an illusion,

(6) series of transfer do not need to be equal to the size of the production series,

(7) sizes of production series do not need to be fixed,

(8) lead time of process is the result of designing process,

(9) when designing a process one should always bear in mind the constraints (bottlenecks).

For successful determination of the flow of process Goldratt proposes the use of network diagrams.

Technological process of production of clothing uses the technique of parallel ways of moving cut parts of garment from one operation to another together with scheduling and deployment of technological operations on the production capacities by checking the availability of resources. Thereby the making or installation of certain parts of garment is performed by an arbitrary number of operations whose scheduling is known to everybody, as well as the duration of individual operations. The optimization of technological operations is shown on Figure 5.6.

The most frequently used expressions for the network activities are CPM

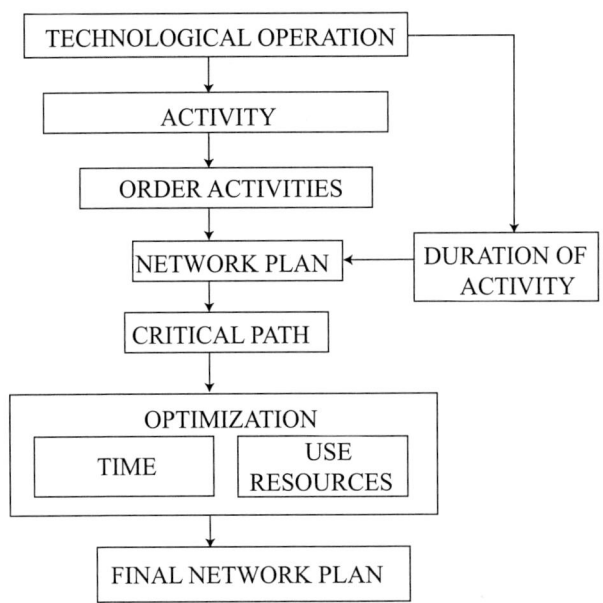

5.6 Optimization of technological operations.

(Circle Plan Methods), PDM (Precedence Diagramming Method) and PERT (Program Evaluation and Review Technique), and the types of networks are:

o AOA (Activity-on-arc).
o AON (Activity-on-node).

CPM for making operational plan is based on the graphical view of activities in the sewing room and the finishing, and their interdependence. This method is used to perform the assessment of duration of individual activities and calculating the earliest and the latest start and the earliest and the latest finish of activities. The earliest and the latest times that overlap are critical events, and provide a critical time.

The advantages of this method are:

❖ previous study of work and process,
❖ identification tasks,
❖ time and material savings,
❖ ensuring the success of the assessment process and control,
❖ optimal layout and
❖ calculating of time reserves for the capacity analysis.

In the following network diagram, activities are represented by oriented lines that begin and end with the identification points, which indicate the events shown in the online rounds, the Figure 5.7.

In large serial production a parallel transition method is applied, because parts of series which are completed in one operation do not wait to for the whole series to end but are sent to the second operation, while making in the first operation is still in progress. This method shortens the production cycle.

Cycle time can be calculated from the formula:

$$T_{cp} = (n-1) t_{omax} + \sum_{i=1}^{k} t_{oi} \quad [\text{hour/cycle}] \qquad [5.12]$$

5.7 Network diagram.

Where: t_{oi} – duration of certain operations,

t_{max} – the longest time in one operation for one piece in series,

n – number of pieces in the series.

In order to avoid delays it is necessary to shift the operations that are on the critical path according to:

$$Ti + 1 < Ti \qquad [5.13]$$

That is, the beginning of production of such operations can move for:

$$\sigma_{1=2} = (n-1)(T_1 - T_2) \qquad [5.14]$$

The mathematical model can be defined as follows:

- Minimum production costs

$$F(x) = MIN\left\{\sum_{i-1}^{ii}\sum_{j-1}^{jj}\sum_{k-1}^{kk}\sum_{l-1}^{ll}\left[(x_1 \cdot x_2 \cdot t_{l,m,j,k}) \cdot c_k + (x_1 \cdot x_3 \cdot t_{i,m,j,l}) \cdot c_l\right]\right\}$$

$$[5.15]$$

$$x_1 \begin{cases} 1 \text{ for selected m variant} \\ 0 \text{ for another variant} \end{cases} \quad x_2 = \begin{cases} 1 \text{ for k} > 0 \\ 0 \text{ for k} = 0 \end{cases} \quad x_3 = \begin{cases} 1 \text{ for l} > 0 \\ 0 \text{ for l} = 0 \end{cases}$$

If is $x_2 = 1$ then $x_3 = 0$ and vice versa.

Restrictions are as follows:

- Taken capacity must be less than or equal to the available one for each capacity

$$ZAK_{kt} \leq RAS_{kt} \qquad [5.16]$$

- All the technological operations must be deployed in the field

$$\sum_{i-1}^{ii}\sum_{j-1}^{jj}\sum_{k-1}^{kk}\sum_{l-1}^{ll}(x_1 \cdot x_2 \cdot t_{i,m,j,k} + x_1 \cdot x_3 \cdot t_{i,m,j,l}) = \sum_{k-1}^{kk}\sum_{t-1}^{tt}ZAK_{kl} \qquad [5.17]$$

- Deadlines in product delivery must be met

$$MAX_{i}(tz_{i,mj-jj,k}) \leq v\,z_n$$

$$[5.18]$$

- Terms of beginnings of first activities must be larger than the system (current) date of drafting

$$MIN_i(tp_{i,mj-1,k}) > s_t \qquad [5.19]$$

Time related to the event (node):

- EET (early event time) – the earliest time when the activity coming from the node can start.
- LET (late event time) – the latest time when the activity entering the node has to finish.

Times related to activity (branch):

- ESij (earliest start) – the EET of node (node i) from which the activity outputs.
- LFij (latest finish) – the FLIGHT of node (node j) into which the activity enters.
- LSij (latest start) – LFij minus the duration of activity.
- EFij (earliest finish) – ESij plus the duration of activity.

TF (total float) is the number of time units for which activity can be moved without moving the end date. Moving the start of the activity may cause postponement of some activities that follow, but its duration will not be jeopardized.

Total float can be initial and final:

$$TF = (LF - D) - ES = LS - ES = \text{initial float time}$$
$$TF = LF - (EC + D) = LF - EF = \text{final float time}$$
$$\text{Or } TFij = LETj - EETi - Dij$$

Where: D – duration

TF is used to determine priorities depending on the backup time that is associated with activities. Leisure spare time is used to determine activities that can be postponed without affecting the total backup time of the activities that follow (Figure 5.8).

$$FF \text{ (Free Float) is } FFij = EETj - EETi - Dij$$

Independent float is used to determine activities that, although delayed, will

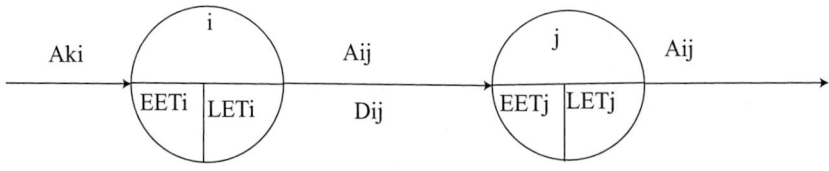

5.8 Free float time.

not affect the total float time of the activities that precede or follow her behind. The value of an independent float time can be negative, and in these cases it is said to have zero value and then both free and independent backup time are the same and equal to zero.

$$\text{IF (Independent float): } IF_{ij} = EET_j - LET_i - D_{ij}$$
$$\text{CF (Conditional float): } CF_{ij} = TF_{ij} - FF_{ij} = LET_j - EET_j$$

The chain of activities that has the longest finish time determines the earliest finish time. This time is often called the time of the project (project time) or duration of the project (project duration) or the most critical path (critical path). Critical time begins with the first node (event) and continues along the network until the final node.

Total float time shares all activities within a single chain of activities. If one activity uses a part of the total float time, the total float time which remains for the other activities is reduced to that amount. Free float time is shared only with activities that proceed and is used during planning. When the float time that some activity has is used, then it can become critical, which results in a new critical path. Duration of each activity:

$$D_{ij} = (Q \bullet nv) / (N \bullet t) \text{ [days]} \qquad [5.20]$$

Where: Q – quantity of some activities on the basis of technical documentation,
  N – number of workers/ working groups,
  nv – standard time unit of product,
  t – the duration of shifts/working hours.

PDM (Precedence Diagramming Method) – has the following relations between activities:

o FS (Finish to Start) – the beginning of the next activity is after the end of the previous one.
o SS (Start to Start) – the beginning of the next activity is after the beginning of the previous one.
o FF (Finish to Finish) – the completion of the following activity is after the completion of the previous one.
o SF (Start to Finish) – the completion of the following activity is after the beginning of the previous one and therefore it is rarely used.

To pass through the network in advance:

$$RP_j = \max (RP_i + SS_{ij}; RZ + FS_{ij})$$
$$RZ_j = \max (RZ_i + FF_{ij}; RP_i + SF_{ij}; RP_j + D_j) \qquad [5.21]$$

To pass back through the network:

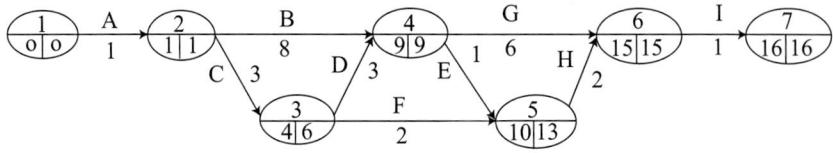

5.9 Analysis of the critical path.

$$KZi = \min (KZj - FFij; KPj - FSij)$$
$$KPi = \min (KPj - SSij; KZj - SFij; KZi - Di) \qquad [5.22]$$

Analysis of the critical path (Figure 5.9):

(1)   earliest start and latest start time must be equal ES = LS
(2)   latest finish and finish start time must be equal EF = LF
(3)   duration of activity is the difference between latest finish and earliest start activities of LF-EC = D

For example, time of making men's shirt is 27 min. Time of making the front of the shirt is 3 minutes and it is a critical path. To minimize the transportation, the production is planned to take place in bundles of 5 pieces. On the bases of the time division of operations per working places for each operation, a longer period of time for obtaining the time of performing each operation that is written in the network plan is taken and the total time for the first bundle is calculated.

For example: $T_{C1} = 0.0311 \cdot 5 \text{ pieces} = 0555$

$$T_{C2} = 0.0069 \cdot 5 \text{ pieces} = 0345 \text{ etc.}$$

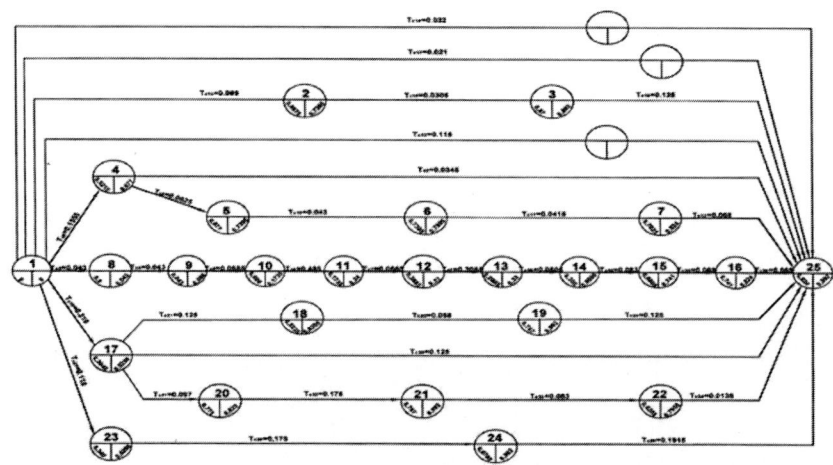

5.10 Network diagram for manufacturing shirt.

A network diagram (Figure 5.10), which was made according to the operation plan for shirts, can help to determine the critical technological operations:

Table 5.2 Analysis of float time

Activity	D	ES	LS	EF	LF	CF	FF	IF
1-2	0.069	0	0.6675	0,.69	0.7365	0.6675	0	0
1-4	0.1555	0	0.5215	0.1555	0.677	0.5215	0	0
1-8	0.043	0	0	0.043	0.043	0	0	0
1-17	0.219	0	0.3045	0.219	0.5235	0.3045	0	0
1-23	0.1805	0	0.345	0.1805	0.5255	0.345	0	0
1-25	0.115	0	0.777	0.115	0.892	0.777	0.777	0.777
1-25	0.201	0	0.691	0.201	0.892	0.691	0.691	0.691
1-25	0.022	0	0.87	0.022	0.892	0.87	0.87	0.87
2-3	0.305	0.069	0.7365	0.0995	0.767	0.6675	0	0
3-25	0.25	0.0995	0.767	0.2245	0.892	0.6675	0.6675	0
4-5	0.625	0.1555	0.677	0.218	0.7395	0.5215	0	0
4-25	0.345	0.1555	0.8575	0.19	0.892	0.702	0.702	0.1805
5-6	0.043	0.218	0.7395	0.261	0.7825	0.5215	0	0
6-7	0.0415	0.261	0.7825	0.3025	0.824	0.5215	0	0
7-25	0.068	0.3025	0.824	0.3705	0.892	0.5215	0.5215	0
8-9	0.043	0.043	0.043	0.086	0.086	0	0	0
9-10	0.0665	0.086	0.086	0.1735	0.1735	0	0	0
10-11	0.0485	0.1735	0.1735	0.24	0.24	0	0	0
11-12	0.0665	0.24	0.24	0.2885	0.2885	0	0	0
12-13	0.3055	0.2885	0.2885	0.33	0.33	0	0	0
13-14	0.0805	0.355	0.355	0.6605	0.6605	0	0	0
14-15	0.083	0.6605	0.6605	0.741	0.741	0	0	0
15-16	0.068	0.741	0.741	0.824	0.824	0	0	0
16-25	0.068	0.824	0.824	0.892	0.892	0	0	0
22-23	0.097	0.219	0.641	0.287	0.709	0.422	0	0
17-18	0.125	0.219	0.5235	0.316	0.6205	0.3045	0.548	0.2435
18-19	0.058	0.219	0.767	0.344	0.892	0.548	0	0
19-25	0.125	0.287	0.709	0.345	0.767	0.422	0.422	0
20-21	0.175	0.345	0.767	0.47	0.892	0.422	0	0
21-22	0.083	0.316	0.6205	0.491	0.7955	0.3045	0	0
22-25	0.0135	0.491	0.7955	0.574	0.8785	0.3045	0.3045	0
23-24	0.175	0.574	0.8785	0.5875	0.892	0.3045	0	0
24-25	0.1915	0.1805	0.5255	0.3355	0.7005	0.345	0.345	0

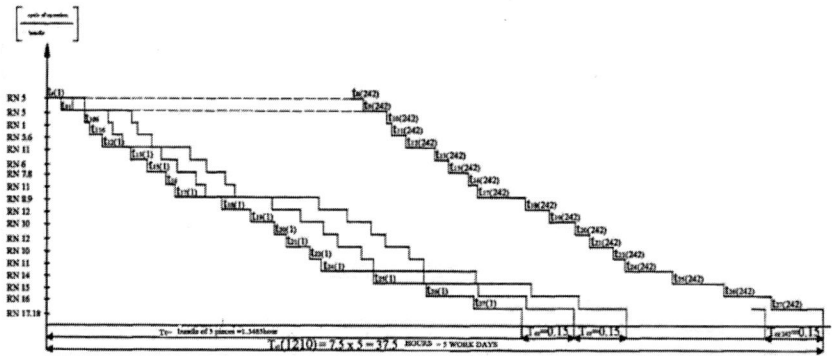

5.11 Term action plan on the critical path for the production of shirts.

1-8-9-10-11-12-13-14-15-16-25

According to the network diagram a critical path is the one at which the backup time is zero. To calculate the float time (Table 5.2) the duration time of activity is observed.

Considering the fact that the technique of parallel ways of moving material from one operation to another is applied, a schedule of production can be suggested only for activities which are on the critical path. This plan (Figure 5.11) shows that the first bundle ends in 0.892 hours. The production of each following bundle ends in 0.3055 hours, since the cycle length of the longest operation is 0.3055 hours for one party. For the week of five working days (37.5 h), a possible production of 598 pieces can be determined in the following way:

- First day $5 + \dfrac{7.5 - 0.892}{0.3055} \cdot 5 = 110$ [pieces/day]

- Second day $\dfrac{7.5 - 0.892}{0.3055} \cdot 5 = 122$ [pieces/day], and so on.

Delays can be avoided by shifting operations that are on the critical path and the start of production of these operations may move due to the continuity of production:

$$\sigma_{5-6} = 597 \, (0.0875 - 0.0665) = 12.537 \text{ [hour/cycle]}$$
$$\sigma_{6-7} = 597 \, (0.0665 - 0.485) = 10.746 \text{ [hour/cycle]}$$
$$\sigma_{20-24} = 597 \, (0.3055 - 0.0805) = 134{,}325 \text{ [hour/cycle]}$$
$$\sigma_{25-26} = 597 \, (0.083 - 0{,}068) = 8{,}955 \text{ [hour/cycle]}$$

Therefore, if the cycle of first operation of first party starts at 7 h, then the cycle of second operation of the first party starts at 7 ^{05} h.

Application of network diagram allows performing of the time calculation of network diagram, the optimal allocation of resources and seeking the best economic solution for production. Network diagram in the technical-technological preparation of garment production allows:

- ❖ a reviewed plan of making a garment,
- ❖ a logical scheduling of technological operations,
- ❖ optimal preparation and making of clothing,
- ❖ an easy defining of deadline of clothing by determining deadlines of certain activities,
- ❖ an easy identification of activities, whose shortening reduces the total time of making products,
- ❖ informational data processing, etc.

PERT method of evaluation and auditing is a technique of network analysis which is used to estimate the duration of the operation when there is a high degree of uncertainty of the duration of individual activities, or when the duration of individual activities is not known. PERT uses the average of the distribution (expected value) of time estimation.

In the PERT technique each of the activities require three estimates of activities (Figure 5.12):

- Optimistic time is the duration of activities which can only happen during its performance under very favourable conditions and circumstances.
- Most probable time is the duration of activities that has the biggest individual probability to happen in practice.

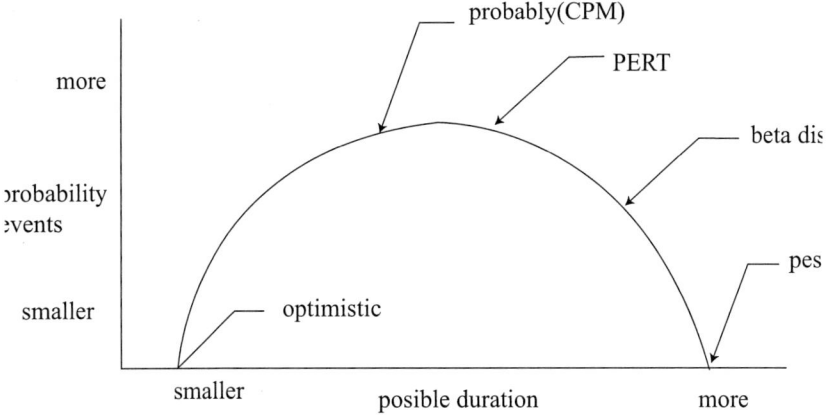

5.12 Calculation of the PERT duration for one activity.

- Pessimistic time is the duration of activities that can happen only under very unfavourable circumstances of performing certain activities.

In PERT method the expected time is:

$$t_e = \frac{a + 4 \cdot m + b}{6} \quad [\text{minute}] \qquad [5.23]$$

Where: a [min] – optimistic (shortest) time,
   b [min] – pessimistic (longest) time,
   m [min] – most probable duration time.

The main difference between CPM and PERT method is that the result of CPM method is the scheduling of activities which minimizes the costs, whereas the result of PERT method is the scheduling of activities which optimizes its duration. The main advantage of PERT method is that it indicates the risks which are connected to the estimation of duration.

Modern information systems (Primavera, MAX for Windows) can be used for a more flexible production planning and monitoring. Those are efficient software packages with programs for production planning, technical documentation, network development plan (CPM and PERT) to 10000 activities, defining all the elements of activities, resource allocation, cost control, reporting system, defining the calendar of production and resources, defining overtime, defining the WBS activities (the Work Breakdown Structure – production oriented tree that leads to the identification of activities) and so on.

In order to achieve optimal production it is necessary to carry out planning of necessary quantity of materials. The amount of material necessary for making a garment in certain size is a normative of consumption of materials. The normative of material consumption is determined on the basis of the length and width of cutting layout, for basic material and extra material separately (lining, interlining). The example of normative consumption for men's shirt is shown in Table 5.3.

Planning of material consumption doesn't require consumptions for each garment size, but it requires calculating of the average material consumption for the group of sizes. The average consumption of materials can be calculated on the basis of absolute or relative participations of individual sizes in the total quantity of products that should be produced in certain period of time.

Based on the relative participation of individual size, the average consumption of material is calculated by the formula:

$$M_{pr} = \frac{M_1 \cdot U_1 + M_2 \cdot U_2 + \ldots + M_n \cdot U_n}{100} \qquad [5.24]$$

Or

Table 5.3 Normative consumption for men shirt

Number	Material	Colour	Dimensions	Unit	Quantity
1.	Cotton fabric	white	1.48	m	1.53
2.	Threads 80/2	grey		m	180
3.	Brand labels			piece	1
4.	Label size			piece	1
5.	Etiquette for the raw material composition			piece	1
6.	Declaration of paper			piece	1
7.	Cardboard labels			piece	1
9.	Cardboard box packing		29cm x 35cm	piece	1
10.	PVC Bag			piece	1
11.	Transport box			piece	0.02

$$M_{pr} = \frac{\sum_{i=1}^{n} M_i U_i}{100} \qquad [5.25]$$

Where: M_{pr} – the average consumption of materials based on the relative participation of certain size,

$M_1, M_2, ..., M_n$ – individual consumption in individual materials,

$U_1, U_2,...,U_n$ – the participation of some size in percent.

On the basis of absolute size of the average share of individual consumption of material is calculated by the formula:

$$M_{pa} = \frac{M_1 \cdot K_1 + M_2 \cdot K_2 + M_3 \cdot K_3 + ... + M_n \cdot K_n}{K_u} = \frac{\sum_{i=1}^{n} M_i \cdot K_i}{K_u} \qquad [5.26]$$

Where: M_{pa} – the average consumption of materials based on absolute participation of certain size,

$K_1, K_2,...K_n$ – the amount of clothing by size,

K_u – the total amount of clothing.

If you move from one width of material to another, then the theoretical consumption of material can be determined by the formula:

$$M_{pt} = \frac{S_1 \cdot M_p}{S_2} \qquad [5.27]$$

Where: M_{pt} – the average consumption of materials,
S_1 – the original width of material,
S_2 – the new width of material,
M_p – the average consumption of material on the width of S_1.

When the length of cutting layout for a particular type of textile material is known, it is necessary to determine, somehow, such way of length arrangement of textile material into cutting layers that the rest of the unarranged length remains minimal, or not to have it at all. The pieces of textile materials (fabrics, knitted) which come to the PBS are of different lengths. As a consequence of different dimensions of material while cutting there comes the creation of increased amounts of waste.

In order to reduce these losses in material, an optimum number of layers in the textile deposits of different lengths which are formed before cutting needs to be planned. The calculation of the number of layers in the textile deposits of different lengths is done by (Trajkovic 1997):

- calculation methods,
- homograph methods and
- computer methods.

(1) The essence of calculation method of arranging textile winding onto cutting layers of different lengths lies in the fact that calculation is used to determine the minimum length of the remainder of the retained length (l_o) of the total length of textile winding:

$$L = l_1 k_1 + l_2 k_2 + l_3 k_3 + \ldots + l_n k_n + l_o \qquad [5.28]$$

Where: L – the length of the winding material,
$l_1, l_2, \ldots l_n$ – the length of individual cutting layers,
$k_1, k_2, \ldots k_n$ – the number of layers in individual layers.

If the length of material in the textile winding is arranged in a single deposit of length (l_1), calculation would be:

$$L = l_1 k_1 + l_o \qquad [5.29]$$

In the event that the remaining of length (l_o) is too large, then the material is distributed in two or more deposits of different lengths. When calculating with two deposits of length $l_1 < l_2$, the maximum amount of layers must be determined (k_{max}) in a deposit of smaller length:

$$k_{max} = \frac{L}{l_1} \qquad [5.30]$$

The rest of the length of material in textile winding is divided into the difference between length of layers l_2-l_1. Thus, the amount of layers in the second deposit is determined:

$$k_2 = \frac{L - l_1 k_{max}}{l_2 - l_1}$$ [5.31]

The amount of layers in the first deposit (k_1) is determined by the formula:

$$k_{2 = } k_{max - } k_2$$ [5.32]

The rest of the length of retained material from the textile coil is calculated by formula:

$$l_0 = L - l_1 k_1 - l_2 k_2$$ [5.33]

(2) Homograph method of arranging length of textile windings on cutting layers of different lengths consists of the replacement of calculations using equations with graphic operations. For this purpose homographs are constructed on the basis of which it is later necessary to find out a combination of arranging length on textile windings on cutting layers without the remaining material or with the minimal one.

$$L = l_1 k_1 + l_2 k_2 + l_3 k_3 + \ldots + l_n k_n + l_0$$ [5.34]

Where: L – the length of material in winding,
$l_1, l_2, l_3, \ldots l_n$ – the lengths of individual cutting layers,
l_0 – the remaining of unarranged length of winding.

(3) Computer method of arranging length of textile windings on cutting layers relies on the possibilities provided by computers in the process of fast calculation and finding out the optimal values of arranging length of textile winding on cutting layers textile without the remaining material. Aiming to arrange the textile winding length rationally, the computer solves the following formulas one after another:

$$L = l_i k_i + l_{o1}$$ [5.35]

$$L = l_1 k_2 + l_q k_3 + l_{o2}$$ [5.36]

$$L = l_1 k_4 + l_q k_5 + l_r k_6 + l_{o3}$$ [5.37]

Where: l_1, l_q, l_r – length of cutting layers in m,
$k_1, k_2, k_3, k_4, k_5, k_6$ – the number of cutting layers in pieces.

Total number of equations is determined by the number of combinations, depending on the default length of individual cutting layers.

$$m = n + \frac{n(n-1)}{1.2} + \frac{n(n-1)(n-2)}{1 \cdot 2 \cdot 3} + \ldots + \frac{n(n-1)(n-2)(n-3)\ldots[n-(n-1)]}{1 \cdot 2 \cdot 3 \cdot 4 \cdot \ldots \cdot n}$$

[5.38]

Where: m – number of equations,

n – types of cutting layers.

In addition to these methods there are other computer methods of rational arranging of length of textile winding on cutting layers of different lengths. The computer provides a number of different solutions of arranging length of material depending on the length of cutting layout or cutting layers.

5.5   Layout optimization

The choice of equipment is done during the establishment of PBS, a reconstruction or modernization. The choice of equipment depends on: sort, type and quantity of production, price, possibilities to achieve savings in time or work, precision, mode and level of work, repair and maintenance costs, depreciation costs, etc.

Types of equipment include:

- equipment that is spent in the immediate process of production (machines for cutting, machines for cutting layers, ordinary sewing machines, special machines, automatic ones, etc.).
- equipment for processing and transmission of energy,
- control and measuring instruments,
- transport devices and
- other equipment (computers, inventory, etc.).

Spatial layout of equipment should allow most rational movement of objects of work in the business process, i.e. the shortest possible and straight. Layout depends on:

- precise definition of the production process,
- elaborate technological process of making a garment,
- types of equipment,
- available space and surfaces and
- layout of buildings.

In Figure 5.13 there is the spatial layout of the ground floor of a manufacturer of women's jackets with workplace layers and in Figure 5.14 there is the spatial layout of the first floor.

There are three classic layouts:

(1)   Line layout – machines are set according to the flow of the manufacturing operations (the object of work is moving from one machine to another in the same direction). It is used in "chain system" of production (when production process is divided into segments – following one after the other, performed in the specified intervals, synchronized and connected

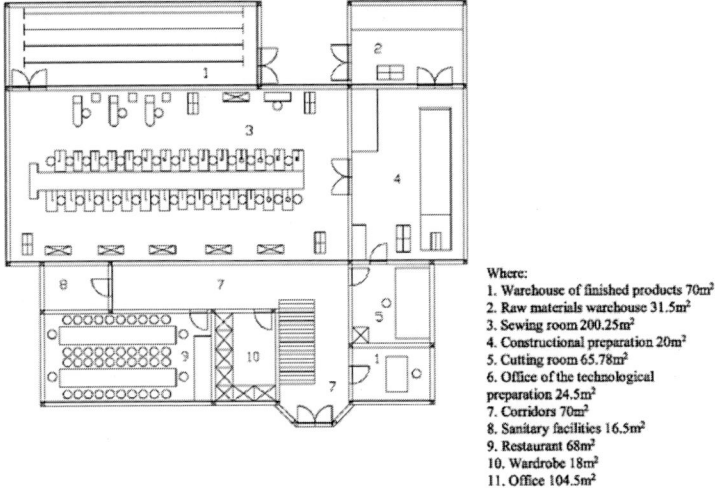

Where:
1. Warehouse of finished products 70m²
2. Raw materials warehouse 31.5m²
3. Sewing room 200.25m²
4. Constructional preparation 20m²
5. Cutting room 65.78m²
6. Office of the technological preparation 24.5m²
7. Corridors 70m²
8. Sanitary facilities 16.5m²
9. Restaurant 68m²
10. Wardrobe 18m²
11. Office 104.5m²

5.13 Spatial layout of the ground floor manufacturer of women's jackets with workplace layers.

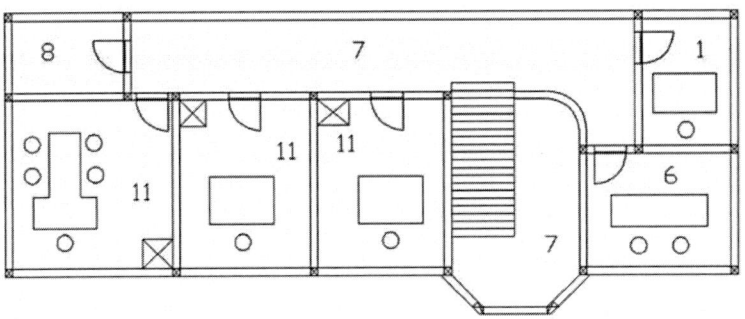

5.14 The spatial layout of the first floor.

in a harmonious whole).

(2) Group layout – machines are grouped according to production – technological features (special machines on one place, automatic ones on another, etc.) and the object of work moves in curves through the production process.

(3) Combined layout – a combination of line and group scheduling, if technological process requires so.

When designing the layout of machines and workplaces in the manufacturing plant there are six questions to be made (Table 5.4) for each event that can happen along the production line, but together with considering the possibilities of their eliminating, combining, changing order, or facilitating

Table 5.4 Questions for the optimal design of the layout of machines and workplaces in manufacture

question	for each	so that we can
why	operation	eliminate
who	transport	combine
what	control	change the order
where	stored	
when		
how much	disposed	facilitate

and/or simplifying the course of each operation, or activity.

The process of planning the layout of equipment:

(1) Analysis of the flow of process for a product.

Schematic diagram of manufacturing operations has special importance for determining the layout of machines, facilities, equipment and internal transport.

(2) Analysis of the flow of products for a number of products.

When analyzing a number of products a combined chart or flow process chart is used. The aim of this analysis is, by combinations of products or product groups, to review the overall flows of materials and semi-finished goods. The basis of the analysis consists of machines and workplaces that are used for the production of various products. Reviewing of their capacity determines their number and studies the performances necessary to ensure the desired requirements in terms of quantity and quality.

Flow process chart of production systematically describes the process or the work cycle and provides enough data for analysis and improving methods of work.

Flow process chart compiles the activities that occur during the process into five groups: operations, transportation, control, hold and store. To describe the event, the following symbols are used (Figure 5.15).

(3) The procedure of application of group technology

A Russian scientist Sokolovski defined types of technological

Symbol	Description
◯	Operation,processing
⇨	Transport,movement
▽	Inventory,warehouse
◗	Retention, temporary postponement of work piece, work piece on waiting setting the container or the infinite strip
▢	Quality control,work on the part
△	The combination of simultaneous course of the operation and control

5.15 Display symbols that are used to map the flowchart process.

processes and that standardization facilitated the process of designing by classification of parts on classes, underclasses, groups and types. By developing this thesis Sergei Petrovich Mitrofanov (1996) started from the possibility of using similar machines and tools for the simultaneous processing of identical operations. Group and standardized technologies made it possible for a broader range of products that fit individual and small serial production to organize large serial production of parts which are common to all products. Standardized technology is defined by a technological process which is unique for all parts that are constructive and technologically similar (men, women and children's trousers, shirts and blouses). Group technology is defined by the unique equipment (sewing machines, steam irons and presses, accessories, etc.) which is necessary to perform an operation on a group of parts that have the same or similar technological operation, but the construction is different.

It is therefore essential, when designing new apparel products, to make the unification of product components (e.g. collars, cuffs, pockets) in shape, size and type of textile material. This unification reduces:

○ types of technological equipment,
○ number, size and types of tools,
○ range of materials,
○ the diversity of production,
○ the stock of tools and materials.

Differences between sizes and materials of similar components should not be large in order not to create losses of materials during cutting or in price during acquisition. The largest numbers of parts (up to 80%) is manufactured in serial production with the line layout of technological equipment, and a smaller part in the singles with the group schedule of workplaces. This is the way to reduce the production time, use capacities better, make transport between machines faster, increase the production volume and the work productivity.

Advantages of group and standardized technologies over classical technology are significant when it comes to manufacturing a wide range for market needs. For the application of group and standardized technology it is necessary to:

– perform the classification, i.e. grouping of parts,
– determine the technological operations for groups of parts,
– design unified and specialized tools and devices,
– modernize technological equipment and
– make the organization of group production lines.

All similar parts of garment are classified as:

o classes – a number of parts which have a common purpose and structural form,
o groups – a set of parts which differ in details and in a number of technological operations and
o types – parts which have the same technological development operations.

The main criteria for the classification of parts are:

(1) similarity of geometric shapes,
(2) similarity of approaches,
(3) similarity of technological equipment used in the making and
(4) similarity of technological process.

Grouping of parts into groups with common technological features is the basis for designing flexible technological modules. Group and standardized technologies are a first scientific concept in production technology.

The same parts within certain technological operations require developing standardized grips and that is how technology becomes unified and technological preparation becomes simple and fast. Studying the layout of equipment and the schedule of grips and movements when performing an operation enables projecting of schedule with a shorter path of moving and optimal order of grips and movements in operation. Better layout and scheduling lead to increasing labour productivity and better humanization of work as well as better utilization of existing resources, shortening the total length of the moving objects of work, reducing the number of procedures, shortening the duration of the operation. It is particularly important to pay attention to the following rules:

- workplace must have an optimal square,
- working conditions should correspond to standards,
- equipment should enable work in a standing or sitting position,
- equipment should be located in the optimal zone which was selected according to the frequency of handling,
- layout of equipment should provide the optimal sequence of movements during operation and
- layout of equipment for supplying the workplace should be optimal in relation to employees and internal transport.

Internal transport involves overall movement of material inputs and finished products in the production process. Rational organization of inner transport is significant because:

- it reduces the time of transport, contributes to faster performing of the process and reduces costs,
- the introduction of automation lead to integration of production and transport into a single production process in which many operations take place during transport (integrated manufacturing systems) and
- the way of performing transport affects the production itself.

Rational organization of inner transport is influenced by:

- layout of premises and workplaces,
- choice of directions of movement and
- choice of transport device.

The task of internal transport is to transfer the right material, deliver it on the right place at the right time, in the required amount, according to the requested order to provide conditions for achieving minimum production costs. The organization of internal transport requires necessary planning and combining of utilization of space, equipment and material that is transmitted in such a way that reduces human labour to minimum and performs it with minimal effort.

Internal transport can be studied from five different dimensions: movement, quantity, time, space and control.

Movement involves the transfer of materials from one point to another. Efficiency of movement is good if it satisfies the factor of safety. The choice of directions of movement depends on the layout of rooms and workplaces, and it must:

- make space and time as short as possible,
- avoid twisting and empty moves and unnecessary stoppings.

Moving can be vertical (overhead conveyors) and horizontal. Horizontal movement can be:

- The system of ordinary relations – cutting parts of clothing (material inputs) are transported to a workplace, and then return to the rule no-load (transport device was 50% used).
- The system of radial relations – a complex variant of ordinary relations – the material is transported to many places and return empty, as a rule (as if to repeat a system of ordinary relations for several times, the utilization is 50%).
- The system of circular relations – workplaces are linked according to the sequence of operations so that the object of work is circularly transported from one workplace to another, ending up at the starting workplace (utilization is > 50%).
- The system of cyclic (circular complex) relations – a complex variant of circular relations (as a system of relations compared to the system of ordinary relations > 50%).

The amount of transported material observed according to the exceeded road dictates the way of transfer and properties of equipment for internal transport, as well as costs of transporting goods per each piece.

Dimension of time determines the speed of movement of material through the manufacture. Scope of work during the production process, re-handling, and the schedule of delivering in time indicate the aspects from which the system of internal transport is studied.

Aspect of space reflects: storage, equipment of funds for internal transport and moving on appropriate roads, as well as scheduling and disposal of materials. Rerouting of movement of material, proper marking, management and handling of inventory are just some aspects of the functioning of control dimension of internal transport. Internal transport is an integral part of the layout in manufacture and it must not be observed separately. Changes in the internal transport system will change the layout of workplaces, machinery, equipment, and vice versa.

The choice of transportation resources depends on:

- type of production and types of products,
- characteristics of transportation resources and
- costs of purchasing and utilization.

According to some research, about 25% of total costs of manufacturing are transport costs. Costs can be divided into direct and indirect.

Direct costs are related to: raw materials and purchased parts, depreciation, labour costs, energy used and so on.

Indirect costs are more difficult to determine, but it does not diminish their impact on business results. They include the following costs:

- wages of workers due to waiting for material,
- reduced level of capacity utilization of funds for work due to delays in transport,
- engagement of means for unfinished production,
- costs for internal transport, including their increase which is the result of poor layout of machines, workplace, equipment and facilities and organization of internal transport,
- utilization and maintenance of the space necessary for the functioning of internal transport (accommodation space for devices and equipment for internal transport, overhead costs associated with space),
- additional costs which are the result of inappropriate handling of transportation resources,
- benefit to customers due to delays in delivery of finished products etc.

It is estimated that internal transport costs make up about 50% of all damages in the industry and 40% to 80% of all operating costs. It can be concluded that characterizes modern industrial production is characterized by, up to now, the highest level of improvement of the organization of internal transport.

Garment industry, for transport of materials from the warehouse of raw materials and objects of work through all phases of production to warehouses of finished products, uses different means of transportation:

➤ in the warehouse of raw materials – trucks, transport trolleys and overhead conveyors,
➤ in cutting room – trucks for transfer of material on pallets, transport trolley for textile coils or cutting parts, overhead conveyors for textile coils or cutting parts and endless conveyor belts on the tables for laying and cutting,
➤ in the sewing room – handling bundles carts, conveyor belts and overhead conveyors,
➤ in the finishing department – handling bundles carts, conveyor belts and overhead conveyors and
➤ in the warehouse of finished products – trucks, transport trolleys and overhead conveyors.

5.6   Logistics in garment industry

The goal of each manufacturer is sales, especially with no storage or pre-production for a customer. It happens more often that the finished products must be held some time in warehouses before delivery, which causes additional costs.

In order to create and maintain the required level of inventory of necessary materials each PBS must know when to produce new amount of product and what that amount is. The time to produce a new quantity of product depends on:

– the length of time of ordering and arrival of primary and auxiliary materials (fabrics, knitted, setup, interlining, thread, buttons, labels, etc.),
– rate of consumption of supplies,
– time of clothing and
– rate of servicing demand.

Determining the quantity of supplies depends on the costs of production and storage costs. The bigger series are produced, the lower costs per unit of material are, but the bigger costs of storage are. In order to determine the optimal level of supplies, the sum of the costs of procurement of materials and storage must be the lowest per unit of product. Therefore, the following models of product supplies are used:

– deterministic models – where the annual demand of product and the annual dynamics of sales are known, and
– stochastic models – where the annual demand of product is not known, but we know the probability of movement demand.

Depending on the coefficient of utilization of the market that is to be taken over, manufacturing capacity and cost structure, the PBS strives to satisfy the demand for the product 70%, 80% or 100%. For example, if the delivery period is 15 days, and the rate of consumption of supplies 10 shirts a day, and the current demand should be satisfied by 100%, the amount of new products must be ordered when the unused amount of stocks reaches 15 • 10 = 150 shirts. If the current demand should be satisfied by 80%, the amount of new products must be ordered when the unused amount of stocks reaches 15 • 0.8 = 120 shirts (product units).

Transport and distribution in the right place at the right time, with the least possible cost, is the aspiration of every manufacturer. Logistics requires the following activities from manufacturers:

● designing and construction of fashion products should take dimensions of the product into account,
● the way, the size and shape of packaging,
● factors of arranging the commercial packaging and its fitting into transport packaging,
● utilization of cargo and storage space,
● manipulative characteristics of products, commercial and transport packaging.

As the increasing number of apparel manufacturers concentrate their production capacities in certain regions due to large investments in machinery,

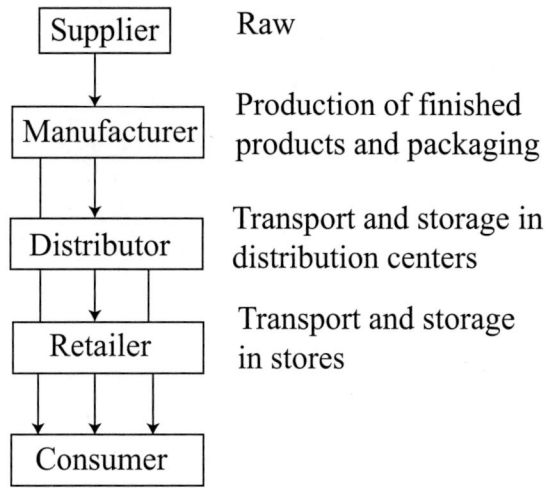

5.16 Supply chain.

the distance between production facilities and customers increases, as well as the appropriate distribution costs. Transport and distribution are a significant business cost that sometimes goes up to 15% of total income in industrialized countries, and in garment industry goes up to 40%.

The question whether to perform transport using its own vehicles or a carrier is very closely linked with the decision about the physical distribution system. If we look at the supply chain (Figure 5.16), manufacturers rarely have wholesale clothing, but they place fashion products into retail outlets. Regardless of the distribution system, while transporting clothes one must pay attention to the raw material composition and the method of packaging.

Packaging of garment and trend mark of product are sometimes more significant for consumers than the quality of product. That is why packaging must be attractive and practical for consumers, distributor and retail. Physical packaging (for example, a shirt put in a transparent plastic bag or a paper box) should not only provide protection and benefits for easier distribution, but also transfer a psychological message to consumers:

- o the packaging indicates the identity of product,
- o the colour gives a necessary image of manufacturers,
- o modern and readable writing style,
- o marks of size and instructions for maintenance, etc.

When an over garment is transported by trucks, it is hung, while shirts, underwear, sports wear, etc. are packed in cardboard boxes, Figure 5.17.

Packing clothes in cardboard boxes allows easy transportation, but clothes packed like that are resistant to rain, molds, insects and sea salt.

5.17 Packing clothes in cardboard boxes or trucks.

In order to secure survival of clothing industry in turbulent environment it is necessary to meet customer requirements with respect to quality, price and delivery term. These criteria can be fulfilled only with the restructuring of existing business and production systems by introducing modern technologies, changing organizational forms, and quality assurance of products during transport to customers.

For a better placement in the domestic and foreign markets it is necessary to observe all the factors of influence on marketing logistics (Table 5.5) and ensure optimal services of clothes' delivery. Major role in the implementation of the logistic task of transport is played by a transporter who is a participant in all phases of the implementation of transportation process, which allows the clothes not to rub, damage or lose its quality.

Standard containers (dimensions 5895 x 2350 mm) are suitable for transporting clothing, but clothing requires a certain way of packaging

Table 5.5 Factors of influence on marketing logistics (according to Pfohl, 2000) in garment industry

Components of services / Features	Time of delivery	Reliability of delivery	Quality Delivery		Flexibility		
			Accuracy	Condition	Modality		Information
					Order	Delivery	
Transport	x	x	x	x	x	x	
Storage Management	ɔ	ɔ	ɔ	ɔ	ɔ	◙	
Packaging	◙	◙	x	ɔ	x	ɔ	
Management of supplies	ɔ	ɔ	ɔ				
Implementation of orders	ɔ	ɔ	ɔ		ɔ		ɔ

Legend: x – major impact ; ɔ – medium impact ; ◙ – little impact

5.18 Closing the protective cover.

(horizontal stacking of clothes) and is not resistant to moisture. Containers for clothing with a lever with hangers and the floor covered with carpet are best for handling all types of clothing because they give maximum protection from creasing, dust and dirt. Each garment should be packed in special protective crowns (Figure 5.18), because the clothes are sensitive to mechanical shocks.

During transport clothes must be under a certain temperature, humidity and ventilation, i.e. under certain climatic conditions.

Some micro organisms and mold may occur under the influence of high temperature, and there is a change in physical features of yarn that exceed the clothes created (the loss in strength, elasticity, electrical conductivity). Clothing made of natural fiber is more sensitive than clothes made of synthetic fibers. Low temperatures reduce the volume of fibers. Convenient transport clothes temperature is from 10 °C to 30 °C, while the optimal temperature is 20 °C ± 5 °C.

Optimal relative humidity for transportation of clothing is from 45% to 70%. The amount of moisture is allowed for clothing made from the wool fibers from 8% to 12%, and from the cotton fibers 7.85% to 8.5%.

Clothing is sensitive to different scents. During transport fabrics easily absorb odours from the microclimate in which they are located. Clothing already has characteristic odours that are incurred through processing and impregnation (especially work clothes) and finishing against the insects, so they often have odours which affects their quality.

References

1. Bruhvviler B (2003), *Risk Management als Fuhrungsaufgabe*, Bern, Stuttgart, Wien
2. Colovic G, Paunovic D and Djordjevic J (2005), 'Useing the tehnically reticular planing technical preparation of production apparels', *5th International Scientific Conference of Production Engineering, RIM 2005. Development and modernization of production*, 759–764
3. Colovic G, Paunovic D and Savanovic G (2007), 'CIM koncept u odevnoj industriji', *33. JUPITER konferencija*, Zlatibor 1.38–1.42
4. Colovic G, Paunovic D and Savanovic G (2008*)*, ' Industrially Custom-Made Clothing', *International Scientific Conference UNITECH 08*, Gabrovo, II-285–290
5. Colovic G, Paunovic D, Savanovic G and Radojevic Z (2008), 'CIM koncept kao preduslov za virtualno projektovanje odevnih proizvoda', 34. JUPITER konferencija, Beograd, 110–114
6. Colovic G, Paunovic D and Savanovic G (2008), 'Logistika u odevnoj industriji', *Operacioni menadzment i evropske integracije*, FON, PKS, Beograd, 243–246
7. Goldratt E M (1997), *Critical Chain*, The North River Publishing Corporation, USA
8. Herroelen W, Leus R and Demeulemeester E (2002), 'Critical chain project scheduling: Do not oversimplify', *Project Management*, 33(4), 48–60
9. Keitsch D (2007), *Risikomanaaement*, Schaffer-Poetschel Verlaa, Stuttaart
10. Kotler P (2006), *Kotler on marketing: how to create, win and dominate marks*, The free press, A Devision of Simon & Schuster Inc.
11. Mayers F E and Stephens M P (2005), *Manufacturing Facilities Design and Material Handling*, Pearson, Prentice Hall, Upper Saddle River, New Jersey
12. Mitrofanov S P (1996), *Scientific principles of group technology*, Boston Spa, Eng. National Lending Library for Science and Technology
13. Pantelic T (2001), *Industrijska logistika*, ICIM, Krusevac
14. Paunovic D (2009), *CAD u konstrukcionoj pripremi odece*, DTM, Beograd
15. Paunovic D and Colovic G (2004), 'Primena racunara u modelovanju odevnih predmeta', Informatika u proizvodnom i poslovnom menadzmentu, IPOM, Doboj, 136–141
16. Paunovic D and Colovic G (2004), 'Modelovanje muske kosulje na CAD sistemu', *Savremne tendencije u proizvodnji i uslugama u nasem drustvu*, IPROM, Beograd, 155-159
17. Paunovic D and Colovic G (2005), 'The Application of SWOT Analysis in Designing of Garments', *Menagement*, No 37, 71–77
18. Paunovic D, Colovic G, Maksimovic N and Maric V (2009), 'Optimization Of Design Work Flow', *International Scientific Conference UNITECH 09*, Gabrovo, s9p96
19. Phillips E.J (1997), *Manufacturing Plant Layot*, Society of Manufacturing Engineers, Dearborn, Michigan
20. Pfohl H (2000), 'Aufbanorganisation of Logistik', *ZfB 50*, 11/12
21. Radojevic Z (1997), *Planiranje i priprema savremene proizvodnje*, Novinsko-izdavacka ustanova Sluzbeni list, Beograd
22. Rogale D, Ujevic D, Rogale S F and Hrastinski M (2000), *Tehnologija proizvodnje odeće sa studijom rada*, Tehnolodki fakultet, Bihac
23. Sajfert Z and Nikolic M (2002), *Proizvodno poslovni sistemi*, Tehnicki fakultet M. Pupin, Zrenjanin
24. Stylios G K and Sotomi J O (2006), 'The principles of intelligent textile and garment manufacturing systems', *Assembly Automation*, Volume 16, MCB University Press
25. Trajkovic C (1997), *Tehnologija izrade odece*, Prosveta, Nis

6
Production management

Abstract: Production of clothing is inevitably transforming into a flexible, agile manufacturing, which aims to track the dynamic changes of fashion. Modern production business system should be capable of designing rapidly, should have the ability for flexible changes because of production of new models of apparel products, the ability for quick adjustment of production capacities, the ability for technology of integration and the production with increased variants of apparel products in required quantities. This can be achieved by modern organization of production, such as Just-in-Time, Toyota Production Systems, Total Quality Management and Lean Production.

Keywords: JIT, TQM, lean production, garment industry.

6.1   Production management

In order to secure the survival of the fashion industry in a turbulent environment, it is necessary to meet the customer requirements with respect to quality, price and delivery term. These criteria can be fulfilled only with the restructuring of existing business and production systems by introducing modern technology, changing forms of organization and participation of motivated workers. Over the past few years new concepts and forms of organization have been developed in Japan, USA and Germany, most of which are simultaneous engineering, lean production, fractal manufacturing, business process reengineering and virtual production.

Today, clothing manufacturers are faced with the request to be flexible and be able to offer a wide variety of products to customers. The leading trend in today's business world is the development of time-based competition. This concept is based on the development of new products and production faster than the competition

OBM (Original brand name manufacturing) is production process which allows placement into domestic and foreign fashion market because it is based on the creation of mark trade and brand, not the "no name" products. It is a great challenge for each garment manufacturer to create a brand, especially for small and medium sized enterprise production business systems which lack the resources and marketing teams that large companies have. On the other hand, sensitivity and rapid changes of the market dictate the pace and "looking for" creating a fashion mark trade with which it can survive and grow in a competitive environment.

In order to survive and be competitive on world markets, manufacturers need to work on:

(1)   production of fashion products with a larger share of added value,
(2)   developing mark trade and creating brand,
(3)   development of distribution channels and
(4)   continuing education and training of professionals.

Production of fashion products with larger share of added value relates to the investment in the production of its own products and investment in the CMT (Cut-Make-Trim), but with a larger share of your own materials, auxiliary equipment and the construction of fashion products. These are tangible and intangible assets, i.e.

- development of new products and clothing materials, production processes or services, a significant improvement of existing products, production processes or services with higher added value (consultancy, testing, research, production samples – prototype),
- investment in the development and acquisition of new technologies – machinery and equipment,
- acquisition of industrial skills (know-how), special knowledge and skills and
- vocational training of workers for new technologies in the processes.

Mark trade is one of the basic characteristics of the product and serves to differentiate one specific product from other, similar ones on the market. Performance of mark trade, as one of the basic characteristics of fashion products, depends on the feelings and opinions of specific product consumer. Due to major changes in fashion trends, developing of fashion mark trade exists in all parts of textile garment chain. The existence of the fashion mark trade brings a number of advantages:

- Achieving greater income.
- Possibility of planning proposals.
- PBS image formation.
- Better correlation between production and sales.
- Greater information and identification.

Mark trade and its strategy became one of the important sources of competitiveness, and the immediate benefits of their application are:

- Separating from competition.
- Increasing the value of the product at consumers.
- Enabling easier and faster launching of new fashion products.

Development of fashion mark trade has to include intangible investments into:

- Creating one's own brand
- Designing
- Prototyping
- Commercialization in one's own production
- Professional training staff

When making a fashion brand it is particularly important to create a basic style and brand strategy, and what features we will give to our fashion product, how we will promote it and invest into it. Branded fashion products on the market, unlike not-branded ones, have the following advantages:

- They are easily recognizable.
- They are demanded more.
- Consumers trust them more.
- Customers are more loyal to them.
- They are more resistant to competition.
- They could be sold at a higher price.

In the fashion industry there is great danger that the brand could become old-fashioned or be overcome by competition. Branding as a continuous process through which more developed types of brands are applied (umbrella, source, range, product) enables increased productivity and sells branded fashion products.

Development of distribution includes finding effective forms of distribution through its own network and trade chains. This refers to the development of its own distribution network and cooperative forms of distribution through which grouped small and medium manufacturers are centralized according to the product characteristics with respect to their complementarily. Besides, it is important to continue with creating alternative forms of distribution, such as franchise, internet and more.

6.2   Flexible manufacturing systems

Development of the world telecommunication systems nowadays allows the increase of sale and purchase of various fashion products all over the world, causing shortening of product life cycle and reducing time to introduce products to the market. On the other hand, there comes the global competition and the market can survive only if they reduce all unnecessary costs and expand the range of production, and consumers are considered individually, not as a statistical average size. Therefore, it is necessary to adjust production to market demands, i.e. to apply a flexible production model or flexible manufacturing system which is capable of adjusting to modern requirements quickly and easily.

The trend of production, as the only answer to mass production, is flexible manufacturing. Flexible manufacturing connects continuous production flows and suspension of production, i.e. uses the advantages of these trends, trying to eliminate the drawbacks of both flows. The aim is to make the flows of materials within production continuous with maximum flexibility of production. This means that the system can promptly respond to the requirements of each customer, without coming to a standstill in production, accumulation of unfinished products, etc. Manufacturers of clothing should apply technological innovations, which will lead to achieving the highest possible level of automation of production. In fact, the goal of automation is not the mass production of large series of products at the lowest possible price, as it used to be, but creating a flexible system that can quickly meet specific customer requirements and which allows easy and rapid reorientation from one type of production to another one. This system is the only one which allows manufacturers to adapt to market conditions effectively not sticking two collections a year, but six or more.

Flexibility is a derived value obtained as the ratio of the number of different products and the size of a series or as a reciprocal value of time required for the preparation of production. There are many models of flexible manufacturing systems, or approaches to measurement flexibility, and some of them are as follows:

(1) According to Professor Dragutin Zelenovic's (Zelenovic, 1986) model of flexibility of production system is a measure of their ability to adjust to his surroundings and the demands of the work process in a given time and given environmental conditions. There are
- Flexibility of structure, which represents the likelihood that the data structure will adapt to environmental conditions, projected work process and disturbances in the process of work. The flexibility of structure includes flexibility of flow structure and flexibility of spatial structure.
- Flexibility of process, which represents the likelihood that the given process of work will adapt to environmental conditions successfully and quickly.

(2) According to D. Bennett (1988) PBS flexibility is a key factor in competitiveness. Traditional performance measurement systems are not enough to comprehend the right way PBS should behave in a more unstable environment. When considering this issue the authors start from the notion of "strategic" flexibility, which consists of: the flexibility of resources (business process, work, supply and delivery system) and two levels of system flexibility.

(3) When considering the flexibility of company P. Bolwijn (1986) distinguishes two basic aspects of flexibility such as: time and changes. When researching the aspects of time Bolwijn suggests the following measures of flexibility: the smaller the duration of the production cycle, the greater the flexibility of the system; a number of hierarchical levels indicates a lower flexibility of the system; if the number of organizational units of production increases, so does the duration of transport and waiting for the production, which causes the lower level of flexibility; and complexity of organization, which is expressed by the complexity of structures and procedures. When considering the aspects of changes, Bolwijn differs the flexibility of financial sub-systems, flexibility of knowledge, procedural flexibility and functional flexibility.

(4) When considering the flexibility of production systems Carl-Henric Nilsson and Hakan Nordahl (Nillson and Nordahl, 1995) start from the input/output analysis and differ the flexibility of system output, the transformation in the system and input into the system.

Flexibility of output from the system is the ability of system to respond to changing requirements of environment, i.e. a constant uncertainty, which refers to product specification, time and quantitative dimension of production. In this sense, there are flexibilities of product variation, product development within the stipulated time, the delivery of parts in the planned deadlines and production volume.

Transformation flexibility of the system is obtained by transforming measures of flexibility in the output into the characteristics of the system, where you get a matrix measure of flexibility/features of the system.

Flexibility of input into the system includes flexibilities of all systems engaged before the transformation: procurement, suppliers and transport.

Flexible technological process of production of clothing reduces development time, reduces costs per unit of product, enables the flow of products and rational usage of the machine park, reduce inventories, increases labour productivity, enables humane, lighter and faster performance, reduces fatigue of workers and increases the quality of products based on the creation of a successful fashion brand and reputation in the market.

In order to make a technological system of production flexible it is necessary to:

- Improve continuously.
- Apply new knowledge and experience in management.
- Apply new knowledge in the field of clothing technology and information systems.
- Design workplace so as to achieve a faster handling of work object for any technological process of making (different models and items).

- Conduct regular employee training.
- Work on the acceptance of change.
- Analyze technological operations and procedures of work and carry out their optimization.
- Design workplaces ergonomically.
- Improve internal transport.
- Apply the techniques of network planning of production.
- Define the quality of each fashion product.
- Tend towards "zero defects" and reduce the warehouse of finished products.
- Achieve the team responsibility of all employees.
- Create recognizable fashion mark trades.

6.3   New methods, tools and techniques of garment production organization

Due to the increased competition and survival in the market, garment manufactures must design production systems that are able to respond to customer demands as soon as possible, minimize production costs and produce a cheap product. New methods of modern organization of production can be Just-in-Time, Toyota Production Systems, Total Quality Management and Lean Production.

JIT (Just-In-Time) is a concept which appeared in the twenties of the 20th century. There are indications that a Ford used it when the ore arrived "exactly on time" to turn into steel for body cars and McDonald's to fry hamburgers. However, in 1970 Toyota showed that it could be applied to all industries.

For the development and progress of each PBS a necessary factor is synchronized and precise work that finally gives a quality product. Quality is what consumers consider most important product characteristics, and it is achieved by introducing quality into production. Quality is based on the JIT system in production process since it provides information on what and how we should work to avoid possible mistakes and shortcomings in the work. JIT system is necessary to improve all aspects of control of both the input of raw materials and the intermediate control, and stimulate the workers to make more effort in performing the job, in order to avoid making mistakes and unnecessary repairs.

All processes of JIT system go through receiving, processing, transmission and using information, it is filled with constant receiving and submitting information.

Aiming to make the effects of JIT system positive, it is necessary to have good relations among people in the organization, staff training, upgrading and expanding the horizons of employees. It is also necessary to modernize production so that each employee understands how important and useful it is. JIT system is a good organization where each worker knows exactly and clearly what he should do. Experience of JIT system helps to understand how large business world systems operate and what is necessary to keep its market position.

JIT is an economic term that represents the strategy of reducing costs in the production of clothing where the calculation helps to achieve less storage of basic textile materials (lining, interlining, thread, buttons, etc.), i.e. raw materials or just avoiding the storage of finished garments. This tends to flexible production for a known customer, without storage, it shortens the time of making clothes, synchronizes the technological process of work and balances capacities.

In order to make JIT method successful in garment industry, many requirements must be met, such as:

(1) Quality.
(2) Minimal inventory.
(3) Reliable and solid cooperation relationship with suppliers.
(4) Suppliers located near the company, with reliable transportation available.
(5) Production size in correlation with demand.
(6) Team work, employees responsible for maintaining their equipment, bosses, coaches and mentors respect their employees and participate in the production process actively.
(7) Customer satisfaction.

The essence of application of JIT concept in the technological process of making clothes is that every activity is directed at the right time and in the right place in the appropriate quantity and quality, without the waste of all forms of work.

Basic characteristics of the concept of JIT in the production of clothing are (Figure 6.1.)

- Market-oriented production – customer requirements determine the system of production. It produces garments according to the market research or for a known customer according to the defined quality of garment.
- No storage production – a direct link between manufacturer and customer.
- Transmission planning in production facilities (for example, through the Kanban) and the usage of simple methods of planning that everyone can understand.

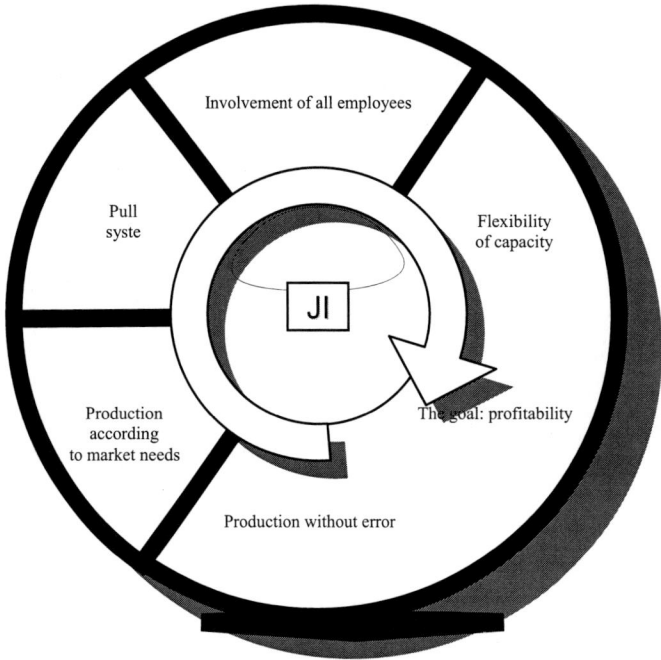

6.1 JIT concept.

- Thinking of production without error, because the manufacturer bears full responsibility for the quality.
- Streamlining of internal transport of materials and flow of work objects from one sewing machine to another.
- Reduction of production cycle – at the same time, the phases of cutting, sewing, and finishing are carried out simultaneously (as soon as part of the planned amount of textile material is cut, bundles are "inserted" into a sewing room and parts of finished garments go into trimming).
- Quality of products and organization of assurance organization. Quality is not what the manufacturer believes is, the quality is what the customer thinks it is.
- Pull-system ("pulls" material into production when required, and reduces the maximum unfinished production and simplifies planning).
- Motivation of employees.

It is very important to engage all employees, or the culture of PBS, i.e. the relation of employees to work. System of "the respect of people" includes

- the application of teamwork,
- extension of the work of employees (for example, every employee is trained to work on three machines – ordinary sewing machine, automaton

and/or special sewing machines) with the inclusion and maintenance and setting machinery and equipment

- encouragement of personal responsibility and a sense of ownership over job.

Therefore, it is necessary to use a basic working method – the method for inclusion in the chain of responsibility:

- Discipline – application of safety standards.
- Flexibility – to remove hierarchical barriers.
- Equality – gender of employees.
- Autonomy – delegating responsibility down.
- Development of employees – increase competitiveness.
- Quality of working life – regulation of working space.
- Creativity – development of jobs.
- Total employee involvement – to give employees more rights, responsibilities and working space.

JIT reveals weaknesses in the technological process of making clothes. Temporary supplies among operations prevent the spreading of interruption of the production process in a technological operation. Real problems are hidden by removing these supplies, the problem spreads quickly, and all the operations in the chain of production stop. It motivates all the people in the chain to solve new problem together, and also to prevent from occurring anywhere in the chain. Therefore, it is necessary to make the control of time making of each technological operation and merge more operations into one, thereby reducing the technological development time, increasing productivity and reducing the price of fashion products.

In the JIT approach in the technological process of making clothes, the layout of workplaces can be according to the Modular Production System (MES). Modular manufacturing is a technological concept used in the Japanese and American automotive industries, which have slowly been introduced into garment industry. Hunter (1990) who defined the modular plant as one made up of many product centres in each of while the complete garment is made by small group of workers. Lowson (1999) produced a comparison of the time taken by traditional bundle systems and team/modular system which suggest that production turnaround time can be reduced from weeks to minutes. In a modular system, processes are grouped into a module instead of being divided into their smallest components.

The modular system was first implemented at company Toyota in 1978 as part of JIT, and was known in the 1980s in the West as the Toyota Sewing System. Monden (1998) gave this system a U-turn layout (Figure 6.2) and

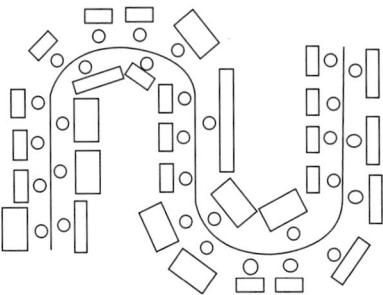

6.2   U-turn layout.

claimed that the main advantage of that system was that the amount produced can easily be arranged by changing the number of operators working in the system. The modular system works on the principle of pull-type production systems, in which the job order comes from the last step to previous steps. Because of this, the amount of work in process is low, even working when no inventory is possible.

The Modular Production System is expressed with team work (up to ten people, then the leader is not necessary, and 20 machines) that allows adjustment to frequent changes in designing clothes, models, small and medium sizes of work order and frequent changes of the order of technological operations. Application of this system leads to reducing of workers' sickness, increasing of production and quality and reducing of downtime in production. Such a system would allow a faster flow of materials (Figure 6.3), rational usage of the existing machine, increased productivity and quality in small batches and allow industrial production to measure (Made to measure).

The Modular Production System allows flexible production of garment, because each workplace can be, for example, equipped with ordinary sewing machines, special machine (for example, overloch machine or special machine with two needles) and steam iron, through a divided system of work where an employee does three technological operations.

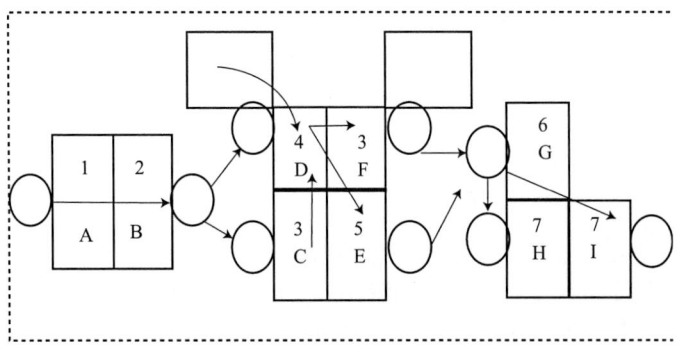

6.3   Flow of materials in the layout modular system.

Adoption of a Modular Production System in garment industry can really bring competitive advantages. That advantage stems from the increased productivity of human labour, reducing supplies of unfinished production, reducing the duration of the production cycle, as well as reducing the space needed for production. On the other hand, adoption of modular system requires profound changes in philosophy and the organization and functioning of production system, since it is necessary to switch from the old, individual method of work to the new, team work, and thereby be governed by the principles of quality, where human resources are the aspect which is of the utmost importance.

6.3.1 Toyota production systems

After the Second World War Japan, faced with prejudices that their products are cheap and of poor quality, introduces the importance of quality and thus creates TQC (Total Quality Control). The basic principle of this concept is that all employees, from managers to workers in production, view a product from the standpoint of consumers. Therefore, quality controls are done with planning and designing with pre-secured quality without waste. It took years to develop special methods of quality management in order to find causes of errors and thereby creating a strategy of quality Zero defect – the concept of American consultant Philip B. Crosby's.

Japanese economy went through a serious rise in the late seventies and early eighties. As a result, the Japanese system of work organization, management and industrial relations has become one of the main subjects of studies of many theorists, and so-called Learn from Japan began to apply, more or less successfully, in other countries.

One of the most successful models for solving problems in production management is the Toyota Production System (TPS). Creator of the Toyota Production System Shingeo Shing (Shing, 1989) defined it as a system that eliminates all the unnecessary things.

TPS is a set of well-known techniques and methods for solving problems, the philosophy about responsible behaviour and returning of values to customers, employees, properties and society. The characteristics of the Toyota production system are

- producing only necessary amounts and types of products within the stipulated time, with the minimum stock,
- employees at all levels participate in the business,
- the system is oriented to the product,
- covers the entire production process to sales of finished products,
- important place is for the quality and costs and

- the heritage of Japanese tradition, history and culture the innovative behaviour of employees is based on.

Important elements of the TPS are

- reduction of a preliminary final time,
- simultaneous operation of workers on more machines,
- quality and
- Kanban system.

6.3.1.1 Kanban

Kanban system is a part of the Japanese manufacturing philosophy that quick and easy methods manage the production. Taiichi Ohno introduced Kanban system in the 50s of the previous century, in order to control the production and introduced JIT at Toyota Company.

Operational planning in its classical sense does not exist, but it is done in manufacturing plants with very short planning horizon. Basic principles of Kanban in a Toyota production system are as follows:

- Continuous improvement of labour productivity, product quality, and efficiency of changing tools, etc.
- Efficient and simple methods for achieving high quality, i.e. the identification of the causes of errors and not finding products with defects.
- Efforts in production process to reduce or eliminate parts that do not contribute to increasing the value of the product (change of tools, transportation, administration, etc.).
- Systematic guidance to reducing the unfinished production
- A reasonable level of inventories of raw materials.
- Bottlenecks in the production process are localized quickly, simply and safely.
- Fast and visual control of the production system.
- Flexibility to market demands.

Kanban is a technique that works on the designing of production plants. Kanban is the signal board of communication about the needs of material and tells the worker about the production of other parts or quantities visually. Kanban system is also understood as a system that "pulls", unlike the traditional "push", parts along the production line – such as JIT or Material Resource Planning (MRP). The problem with the Kanban system is the inability of planning the production. Unlike MRP systems, where the real plan of production is based on previous experience, Kanban depends on the orders and the rhythm of production. In such circumstances the system is more vulnerable to shortages of resources.

6.4 Storage before and after the introduction of Kanban system.

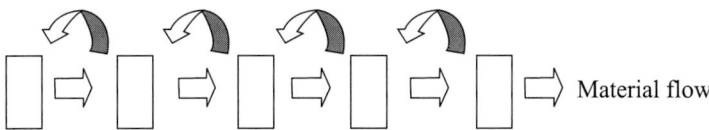

6.5 Production lines in the Kanban system.

The storage of auxiliary materials (thread, buttons, hangers, labels, etc.) before and after the introduction of Kanban system is showed on Figure 6.4.

With Kanban systems work order is delivered to the perpetrator who performs the final operation, and he takes the case from the perpetrators of previous operation, he from the perpetrators of previous operation, etc. (Figure 6.5).

6.3.1.2   PPORF or 20 keys

Program 20 keys or PPORF (The Practical Program of Revolution in Factories and Other Organizations), was developed by a Japanese professor Iwao Kobayashi (1995), and was first implemented in the Toyota Company. It includes 20 practical and integrated methods for improving competitiveness with improving the product, their faster delivery and lower prices. Today, the method is implemented in about 700 companies in 55 countries, which:

- achieve strategic business objectives,
- improve the speed of learning and innovation,
- increase the productivity and flexibility for better acceptance of market demands,
- eliminate all types of waste,
- motivate employees and
- improve competitiveness, profitability and long-term success.

The system includes 20 keys:

Key 1   Cleaning and organization – creating a nice and functional workplace workers are proud of and usage of tools for several different operations which increases productivity.

For example, because of the large number of bales of textile material on the floor next to the machine for cutting and sewing it is necessary to clean and organize the space around them. Workers in cutting room and sewing room must have a clean and ergonomic workplace with special equipment for sewing (for sewing different seams and hem).

Key 2   Rationalization of system – allows the natural flow of the organization of production of various items (shirts, blouses, work clothing) and the development of technological documentation for each product (see Chapter 2).

Key 3   Activities of small groups – using the experience and ideas of all employees in the PBS, or a small group of employees who discuss problems and product quality.

Key 4   Reduce Work in progress – for reducing the amount of finished products in stock.

Key 5   Rapid change of technology – improves flexibility and ability to meet customers' demands and eliminate waste by reducing the time of changes in the workplace. Therefore, it is necessary to conduct market research and marketing analysis.

The new technology for garment production needs flexibility for manufacture in accordance with market requirements. Fast changes in technology as well as customers' expectations make a producer keep improving his fashion products and quality in order to keep his position on the market. Markets researches, consumers' wishes, requests, and criteria mean inevitable and dominant task for a producer of garments, because by obtaining all these information a production can be directed, business planned with advanced defined aims and strategies. Marketing enables greater flexibility and better organization for more successful reaction to market demands.

Market analyses are perhaps difficult procedures for fashion industry, because they need time to see strong sides and opportunities although they are too eager to identify weaknesses and threats. It is important to be aware that once when weaknesses are identified, some steps to change them can be taken by training, so there is possibility to make it a strong side. That's why BSC (Boston Consulting Group), SWOT and Ansoff's matrix are useful techniques used to find out strong and weak points in a fashion industry.

Key 6   Kaizen business (Japanese Kai = change + Zen = good) – real improvements in costs and productivity are achieved and maintained by analysis of procedures that add value, reduce unnecessary movement, combinations and simple procedures. The production of clothes does different garments of different colours and textile materials and that's why each technological operation must be analyzed.

Key 7   Zero defect in the production – the introduction of sewing machines and devices in order to eliminate the control of entire cycle.

Key 8   Related manufacturing – simplification of processes and production lines, reducing and removal of the storage and excessive inventories.

Key 9   Maintenance of machinery and equipment – preventive maintenance of machines for cutting, sewing and finishing (Total Production Maintenance) to increase the performance of the equipment above 95%.

Key 10   Control of production time and the commitment – creating a positive working atmosphere and good scheduling of technological operations.

Key 11   Quality Assurance – clothing quality assurance using CAD/CAM systems and CNC sewing machine with on-line monitor for control of stitches.

Key 12   Developing partnerships with suppliers and customers.

Key 13   Eliminating waste – during the process of technological development, especially in the technological cutting process, there are losses of textile materials, which can be systematized into the following losses: losses in the ends of cutting layouts, losses between patterns (the difference between gross and net areas of cutting layouts), the breadth of material losses, losses as the rest of the textile windings, losses due to material errors, losses due to form and shape of garment, losses due to inadequate number of cuts in cutting layouts and overlapping cutting layouts.

Key 14   Motivation of workers to make improvements and teamwork.

Key 15   The diversity of skills and training of staff training of all employees, training on new processes and using new technologies, but also the training of managers about supporting the staff.

Key 16   Production Schedule – defining the flow of production and modelling process, for example a modular system.

Key 17   Control performances – for example, control the rhythm every two hours in sewing room.

Key 18   Using information technologies – such as applying CAD, CAM and CAP for better preparation and organization of production.

Key 19   Saving energy and materials – the application of CAD/CAM.

Key 20   Using technology for strategic advantage – using new technologies with the benchmarking and development of new fashion products.

Application of PPORF method or system of 20 keys in garment industry allows:

- reducing the cost of procurement and storage of textile materials,
- easier adjusting to new products demanded by the market,
- reduction in staff absence by 14% to 20%,
- continuous flow of material,
- ensuring product quality and
- respecting deadlines.

6.3.2   Total quality management

TQM (Total Quality Management) was created by American professors WE Deming and JM Juran who failed to realize their ideas about the quality of SQC (Statistical Quality Control) in America, but they faced the approval in Japan. After the success of Japanese products in the world market in the eighties, the West became interested in this concept and the concept of TQM appears as a response of the West to the very successful Japanese business concept of Kaizen.

TQM is defined as an approach to quality management in PBS, based on participation of all employees, focused on long-term success by meeting the needs of consumers. TQM is a way to improve the functioning of the PBS continually, on all its levels and using all available resources.

Basic principles of TQM are

- Quality can and must be managed.
- All employees are responsible for the quality.
- Problems must be prevented, not just fixed.
- Quality must be controlled.
- Improving the quality must be a continuous process.
- Objectives are based on customer requirements.
- Management must be involved.
- Improving the quality requires planning and organization.

TQM is applied in order to improve effectiveness, efficiency, flexibility and competitiveness. It requires the entire PBS to be organized and committed to quality in all its segments, in each activity and production unit, so that every employee understands its importance and its commitment.

There are five key components that every PBS must constantly examine and measure in order to realize the level of weakness and productivity. Each of these parts is of strategic importance for the functioning of TQM, and they are connected and dependent on each other. These are as follows:

(1)   Product – meets the requirements of management, all the employees, customers or consumers.

(2) Process – ensures the product quality. Processes must be evaluated to understand if they meet specific standards and expectations of a PBS. Otherwise, the process must be corrected to ensure that the product is of satisfactory quality.

(3) Management team – to ensure the success of TQM it is necessary to have a trained team and the division of responsibilities.

(4) Commitment – researches show that TQM is successfully implemented in the PPS which has a high degree of commitment of workers and management. By their participation in the implementation process, managers can see if the improved processes give the desired results. On the other hand, workers by their assessment in problem solving become motivated to work because their results are recognized.

(5) Organization – an organization can be successful only if there is teamwork. Their creativity, enthusiasm, objectivity and motivation enable functioning of TQM.

Teamwork is a combination of skills, experience and knowledge. A team in TQM provides

- Flexibility, as teams gather, develop, direct and dissolve easily in order to ensure the permanence of structure and process.
- Commitment, because teams are committed to clear goals to which PBS tends.
- Synergic response of teams to challenges, because each complement the skills and experience.
- Motivation, because the work of the individual and his responsibility are important for the team, but the role and performance of each individual have impact on team members.

In order to reach the exact JIT and Quick Response System (QRS) in today's garment production, when planning and specification of quality control garment it is necessary to determine its purpose in accordance with the requirements of customers, to specify the characteristics of products and qualitative values. Quality is the source of rationalization, profitability, competition means, factor of productivity and a precondition for security and creating new jobs.

In dealing with garments it is necessary to satisfy the individual tastes of buyers, design and fashion trends, ensuring individuality and resolve issues of durability, endurance and comfort, because creating a successful fashion mark trade and reputation in the market is based on the quality of garment.

The quality of each clothing production requires

(1)   product quality,
(2)   quality of materials,

(3) quality of process
   ● quality of market research,
   ● quality of design
   ● quality of procurement of materials
   ● production quality and
   ● quality of sales.

Quality of production and material is defined by national standards and the contracts of sale. According to the ISO definition of quality is: "Quality is the set of all the properties and characteristics of products or services related to the ability to meet the established or indirectly expressed needs".

PNQ method (Price Non Quality) or Problem Solving Skills method is a simple method which is based on determining critical errors, determining the corrective measures and implementing corrective measures. This method integrates a number of basic tools for quality improvements such as flowcharts, check sheets, histogram, pareto diagram and ishikawa diagram.

Methods of quality can be divided into three groups:

(1) Basic tools of quality – Histogram, Scatter diagram, Correlation diagram, ABC-Pareto diagrams, Ishikawa diagrams and Control Charts,
(2) Complementary tools of quality – Flowcharts, Nominal group technique, A guide for organizing meetings, Affinity Diagram, Fault tree diagrams, Matrix Diagram, PDPC (Process Decision Program Chart) diagram and
(3) Methods and techniques of quality – brainwriting and brainstorming, SWOT analysis, FTA analysis (Fault Tree Analysis), Value analysis, Network diagram, Kanban, Rolling, Poka-yoke, Zero defect, FMEA method and QFD method.

Methods and techniques of improving quality can also be classified as:

● Statistical methods and techniques,
● Engineering methods and techniques and
● Managerial methods and techniques.

(1) Flowcharts or algorithm allows a simple graphical representation of workflow of the process it examines.
   For example, control design-constructional and technological preparation of production process of cutting can be made by applying the flowchart of the process. Before the technological process of cutting it is necessary to check the: cutting parts (Figure 6.6), cutting layers (Figure 6.7) and cutting layout (Figure 6.8).
(2) Check sheets shows all kinds of errors and their frequency. As an example of Check sheets in Table 6.1 the analysis of commonest errors in design-

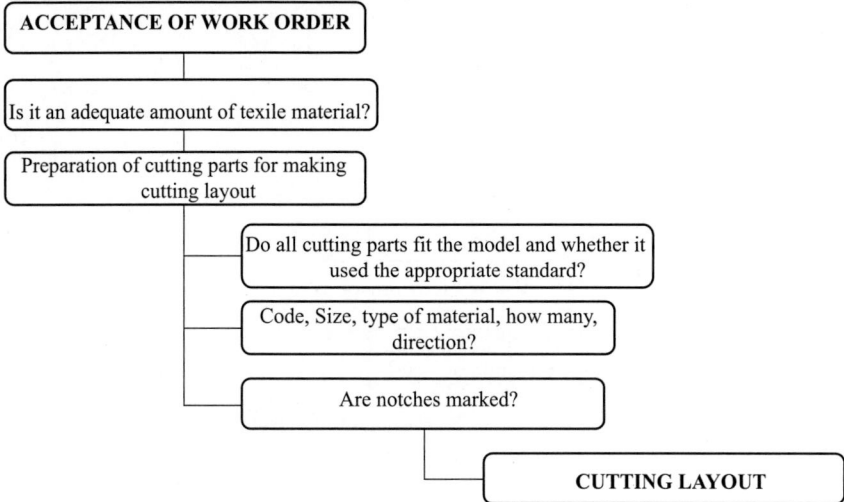

6.6 Flowcharts for control design-constructional preparation.

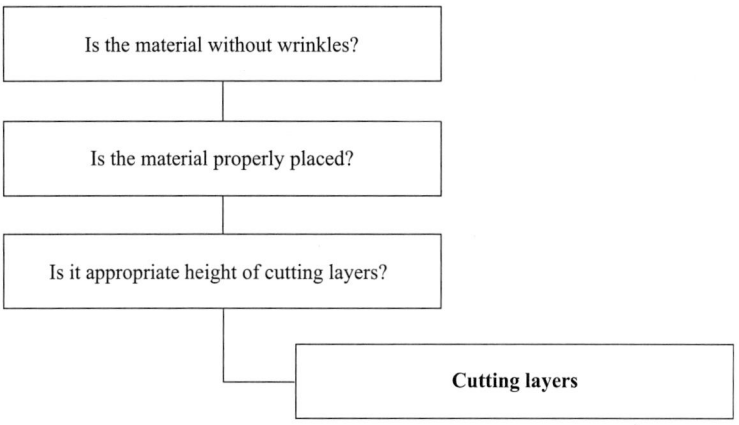

6.7 Flowcharts for control cutting layers.

construction preparation during several months (May to October 2009) is shown.

The technological process of sewing often bears a large number of errors so the intermediate control in sewing room is necessary in order to avoid:

- material damage due to mechanical needle penetration forces, transport, fineness of thread, machine safety,
- drop of penetration parameters due to non-compliance,
- uneven density of stitches,
- unsatisfactory weld strength and elasticity, uneven weld width,
- botch darts, pleat, topstitches,

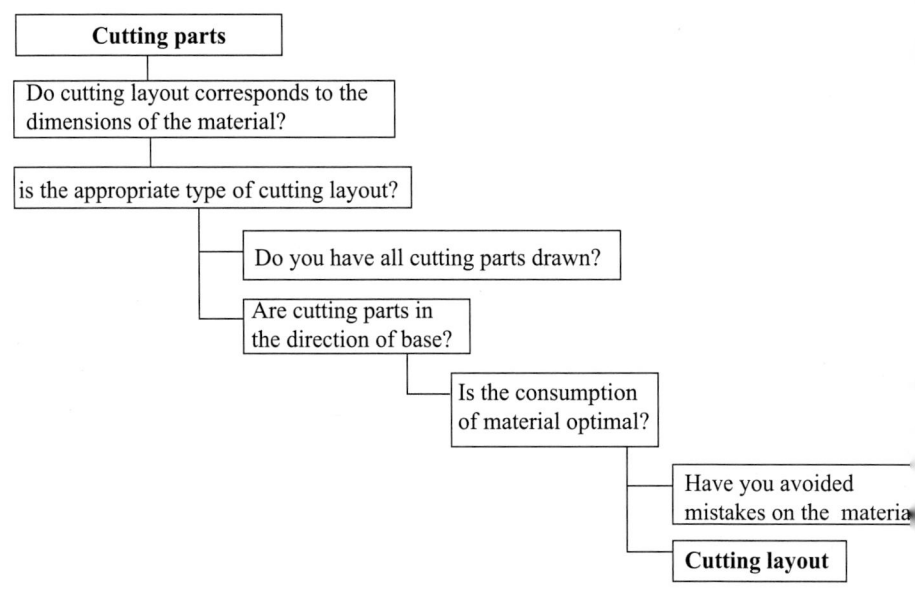

6.8 Flowcharts for control cutting layout.

Table 6.1 Analysis of errors in design-construction preparation

Type of error	Number of errors, fi	Relative frequency,	Cumulative frequency, fk (%)
Cutting parts do not fit the model	2	4.25	4.25
Bad positioning of cutting parts	3	6.38	10.63
Unsigned indentation	10	21.27	31.90
Lack of cutting part	7	14.89	46.79
No percent for stretching	2	4.25	51.04
Deviations in grading	5	10.23	61.27
Bad cut	8	17.02	78.29
No grading all cutting parts	3	6.38	84.67
Large consumption of materials	2	4.25	88.92
Missing cutting part in cutting layout	3	6.38	95.75
Inadequate size of cutting layout	2	4.25	100
Σ	47	100	

Table 6.2 Analysis of errors in the men's shirts

Type of error	Number of errors, fi	Relative frequency,	Cumulative frequency, fk (%)
Weld cracking	4	4.59	4.59
Bad cutting threads on the seam	28	32.10	36.69
Bad consolidation stitches	15	17.24	53.93
Incorrect sewing board	6	6.89	60.82
Bad closed board with topstitch	3	3.44	64.26
Uneven hem	8	9.19	73.45
Uneven making holes	5	5.74	79.19
Unequal length side	8	9.19	88.38
Wrong sewing button	4	4.59	92.97
Stains	3	3.44	96.41
Wrong label	3	3.44	100
Σ	87	100	

- interfusion of cutting parts,
- curvature of sewing parts, asymmetry, spacing, position,
- botch set by size, shape, tension, curvature,
- bad cutting threads, etc.

List of collected errors is displayed in Table 6.2 and they are identified in the production on a random sample of 250 men's shirts.

On the bases of the list of collected errors the degree of preparation can be calculated:

$$P = (\text{number of detected errors}/\text{number of possible errors}) \cdot 100 \quad [6.1]$$

Where: the number of possible errors $= 250$ pieces $\cdot$ 11 elements examined

The degree of preparation of analyzed men's shirts is $P = 3.16\%$

For 250 products surveyed the number of errors is 87, so it can be concluded that every third product does not fit the prescribed requirements of quality.

Figure 6.9 shows the example for flowchart of the process for the selection of adhesive interlining.

(3) Histogram displays the data according to their frequencies. In Figure 6.10 an example of histogram analysis obtained by collecting four samples of t-shirts by width after the first wash at 60°C and 40°C is shown.

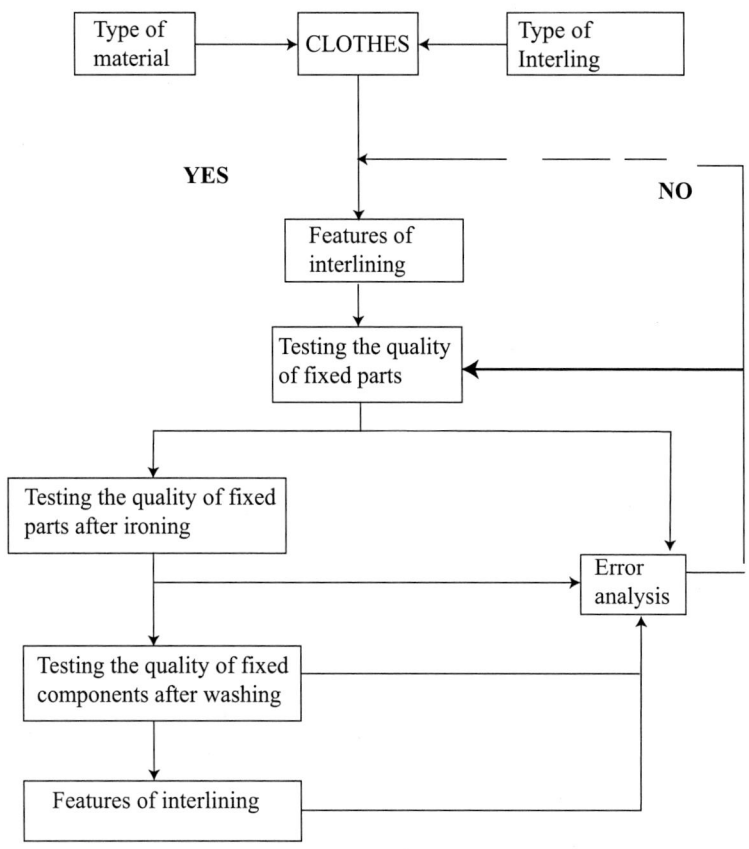

6.9 Flowchart of the process for the selection of interlining.

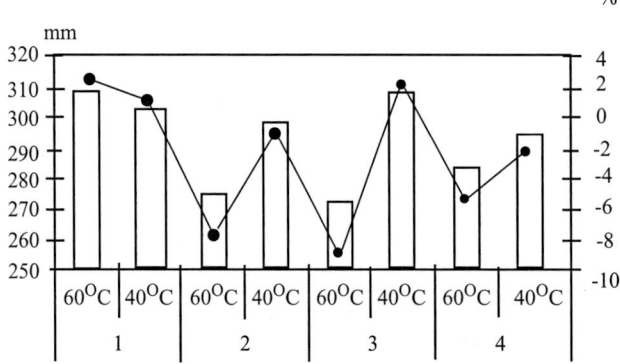

6.10 Collecting samples of four T-shirts by width after the first wash at 60°C and 40°C.

(4) Pareto chart or the ABC diagram allows defining the most important problems and discovering their causes. In 1906 Vilfredo Pareto, Italian economist, sociologist and philosopher defined the rule that 20% of causes are responsible for 80% of errors. Pareto chart provides: ranking according to degree of importance of appearance, determination and separation of the critical areas of the observed size and direction towards solving the problem. Applying Pareto chart in the process of improving the quality of the area includes: management (analysis of sizes of basic indicators of business), marketing (analysis of market trends), development (characteristic trend analysis), production, cash flow management and logistics.

Figure 6.11 shows the example of Pareto chart for the analysis of the production profits.

(5) Ishikawa Diagram, Fishbone Diagram or Cause and Effect Diagram provides a complete analysis of a problem or error, because it observes the environment, methods, material, man and machine.

Ishikava Kaoru (1990) defined that the causes of problems are related to several categories:

- The 5 Ms:
  - Machine.
  - Method.

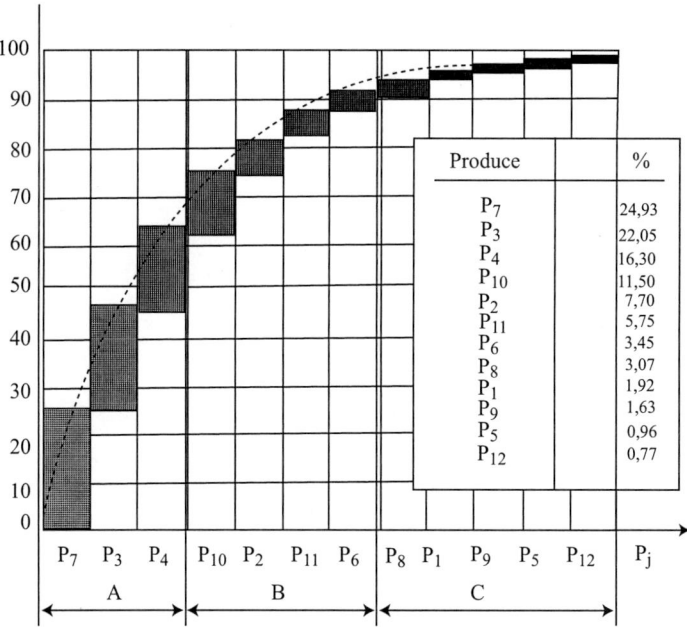

6.11   Pareto chart for the analysis of the production profits.

- ○ Materials.
- ○ Measurement.
- ○ Man.

- The 4 Ss:
  - ○ Surroundings.
  - ○ Suppliers.
  - ○ Systems.
  - ○ Skills.

- The 8 Ps:
  - ○ Price.
  - ○ Promotion.
  - ○ People.
  - ○ Processes.
  - ○ Place/Plant.
  - ○ Policies.
  - ○ Procedures.
  - ○ Product or Service.

Categorize 4 Ss and 8 Ps recommended for service sector.

Figure 6.12 shows the example of Ishikava diagram of cause impact on the quality of clothing

(6) In order to achieve the cause-effect impact for making quality clothing, it is necessary to keep carrying out a Deming PDCA (Plan-Do-Check-Act). The process of continuous improvement is based on the concept which was developed by an American expert on quality William Deming in the fifties of the previous century. In this cycle of continuous quality improvement, shown in Figure 6.13, production can be achieved without deviations and errors, if such was set as a goal.

To make a cycle realized it is necessary to ensure all requirements for quality such as:

- implementation of methods and techniques to detect errors and discrepancies and the reasons for their emergence,
- continuous education and motivation of employees to achieve JIT production and quicker placing on the market (Time to Market)
- quality control as a prevention,
- norms of quality, rather than norms of quantity.

When planning and specification of quality control garment it is necessary to determine its purpose in accordance with the requirements of customers, give precise product features and quantitative values.

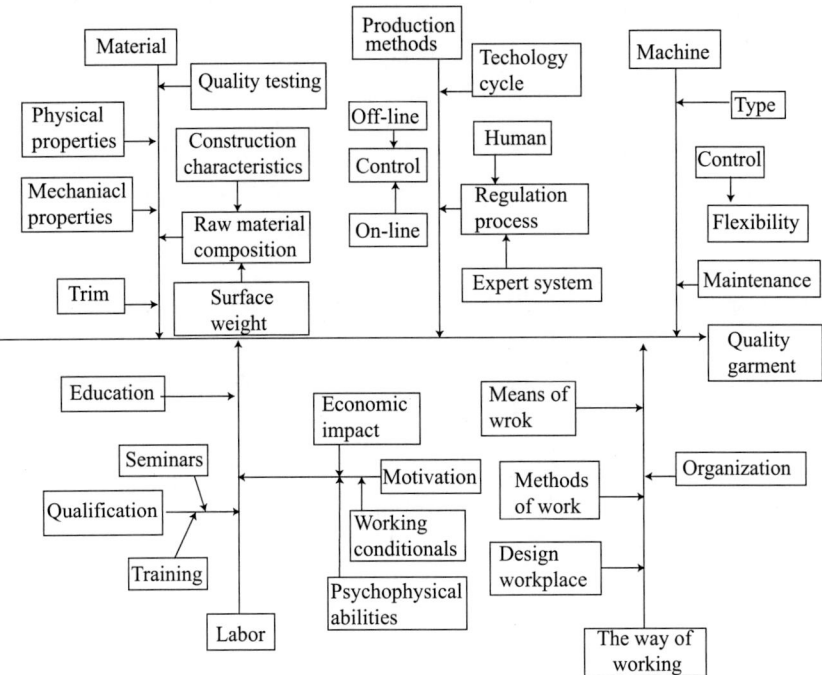

6.12 Ishikava diagram cause impact on quality.

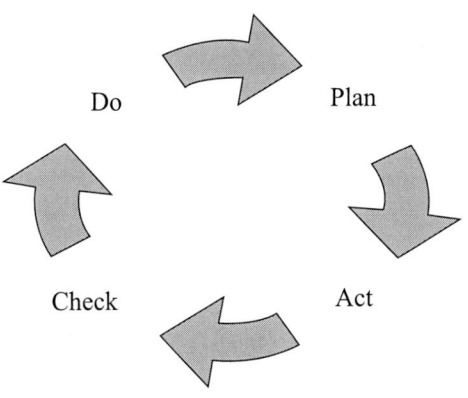

6.13 The PDCA cycle.

Improving quality   →   Fall of costs   → Increased productivity
(minor repairs, defects, delays, damage, better utilization of materials and machines)
Conquering the market with quality → Provided survival → Provided new jobs and lower prices   on the market

6.14 Displaying chain reaction: quality – costs – productivity – the conquest of the market.

Continuous improvement of quality of clothing causes a chain reaction, and inevitably affects productivity, as shown in Figure 6.14.

(7) Poka-Yoke method is the most popular method for ensuring quality without errors. This method of quality control is done during or immediately after a process, not just to define quality in the end. Today, this method is fully implemented in Japan and with the help of auxiliaries (mechatronic devices with stop or sound or light signal) Poka-Yoke detects errors in the object of work and the process stops.

(8) QFD (Quality function deployment) method is mostly used in quality management processes in Japanese companies, and so their experience is being followed in application and in the way of introduction into use. Japanese society for QC recommends QFD for defining "the voice of the customer" (VOC), as a scientific process. The voice of the customer is also the motivating factor for the QFD method, and as such determines success or failure of the product and service on the market. Barnard and Wallace (1994) have integrated QFD as a compulsory method for development strategy defining.

QFD is a powerful tool that enables significant improvements in the product/process characteristics. However, it is not a short-term solution to the product development problems. QFD provides systematic approach to creation of the team outlook, on what needs to be done, the best ways to do it, the best order in which the proposed tasks have to be accomplished and on the staffing and resources that are required to enhance customer's satisfaction. It is also a good format for capturing and recording/documenting the decision making. Applied through the Kaizen philosophy under TQM, QFD is a highly developed form of integrated product and process development in existence. Companies that were using QFD for the product development have experienced, in average:

• 50% reduction in costs,

- 33% reduction in the product development time,
- 200% of increase in productivity.

Based on research Bagozzi (1994) QFD method is based on the VOC. Application of QFD in the development of new methods or improvement of products includes

(1) customer requirements,
(2) product characteristics,
(3) critical parts
(4) critical operations and
(5) steps to be taken.

Implementation of the QFD method in clothing industry is represented through four principles of work in four phases:

- phase 1 – translates customer's demands into the product features,
- phase 2 – translates the product features into features of its parts,
- phase 3 – translates features of parts into technology of fabrication and
- phase 4 – translates technology of fabrication instructions for technical documentation.

On the basis of these principles four houses of quality are formed. As an example the first house of quality was given based on the research of professor Danijela Paunovic (2009) in Serbian garment industry in the Figure 6.15 is phase 1 and establishment of significant characteristics of the garment product.

(9) The FMEA (Failure mode and effects analysis) method performs the identification of all possible errors in the product and the possible risk of their occurrence. The buyer will accept the purchase of fashion products and the overall risk quality installed in that product. FMEA method is a means to identify risks, detect their causes, risk assessment and proposing measures for reducing their occurrence.

The objectives of FMEA method, by definition, are the SMART (Specific, Measurable, Achievable, Related to the customer, Time targeted) objectives for several reasons:

- clearly and unambiguously defined goal of applying FMEA method is the key to success,
- measurable targets can be seen in the pre-defined risk factor priorities which we want to achieve and
- the basic goal of FMEA is to reduce the possibility of errors.

The method is based on defining and determining functions, errors, causes, consequences and other relevant parameters.

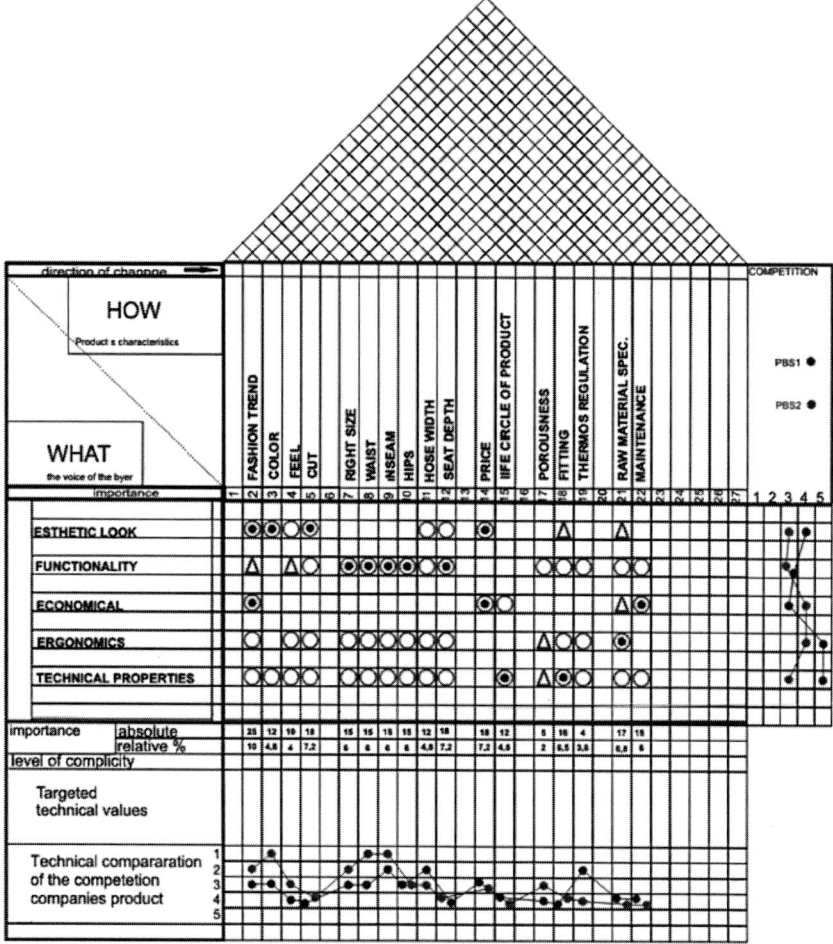

6.15 House of quality – phase 1 and establishment of significant characteristics of the product.

(10) Seven step method provides answers to questions:

- How did we do?
- What to do to improve the quality?

Continuous repetition of seven steps within the PDCA cycle develops many habits (Figure 6.16):

- understanding of the problem,
- development and promotion,
- teamwork and
- diagnostic process.

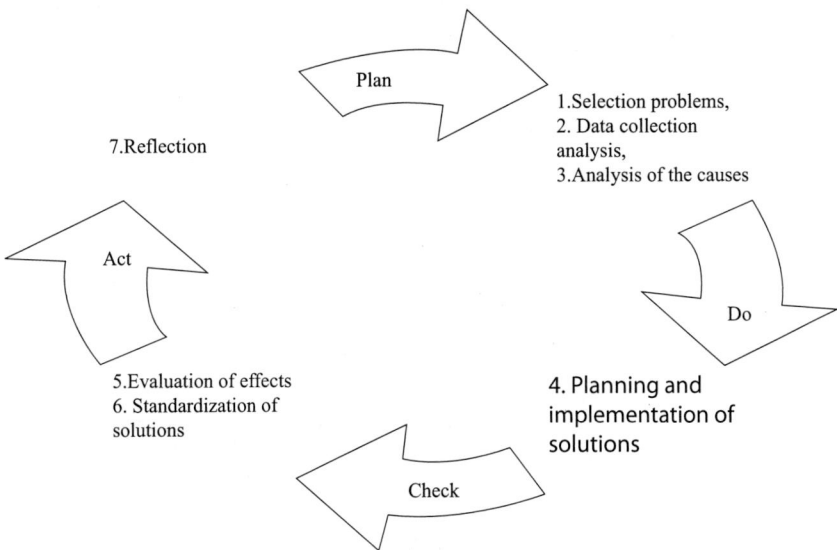

6.16 Method of seven steps within the PDCA.

Table 6.3 shows the most frequently used tools of quality when applying the methods of the seven steps according to Lazic (2005).

(11) Brainstorming and Brainwriting are techniques of generating and collecting ideas: for development of new products, solving management problems, improving product quality and improving sales and marketing products.

Table 6.3 The most frequently used tools of quality when applying the methods of the seven steps

Seven-step	Control Tools	PDCA
Selection problems, data collection and analysis, analysis of the causes	Flowchart; Pareto chart; Histogram; Scatter diagram; Ishikava diagram	Plan
Planning and implementation of solutions	Fault tree diagrams; Matrix Diagram	Do
Evaluation of effects Standardization of solutions;	Cheklsits; Pareto chart; Scatter diagram; Ishikava diagram; Control charts	Check
Reflection		Act

Implementation process of brainstorming involves a few steps:

- – introduction to the rules
- – defining issue,
- – presenting ideas,
- – evaluating ideas,
- – proposing action and
- – adoption of the plan.

Brainstorming is a technique in which a group of six to 12 participants can lead to: complete problem solving, forming a list of possible solutions or a list of ideas that make a plan for finding a final solution.

Brainwriting is a group technique (up to six participants) aimed at the development of ideas, but unlike brainstorming, participants record their ideas. This technique eliminates possible influence of leaders to the opinion of participants, as well as the possible direct or indirect impact on restraining a free and unlimited presentation of ideas.

Japanese industry nowadays develops a separate system of work organization and management, significantly different from the European and American one. It is not only a new organizational form, but also new models of work motivation. A key factor in motivation is reflected in the possibility of creative engagement of all employees through the innovative group. Every worker knows that, if he wants to, he can take part in improving working conditions and production through groups. Quality circles, the production of Zero Defect and timely supply of tape production (JIT) are some of the elements that constitute what is called the art of Japanese management.

6.3.3 Lean production

Lean production (Kaizen techniques originally) is a result of the analysis conducted at the Institute of Massachusetts Institute of Technology (MIT), which was performed for the U.S. automotive industry, with the aim to find a key success of Japanese manufacturers.

Taiichi Ohno, the director of the Japanese company Toyota, was the first who practically implemented the concept of lean production. Lean manufacturing refers not only to production, but also to all other functions within the business systems, as well as relationships with manufacturers. Thus organized PBS through the centralization of technical-economic functions allow:

- reducing wearing of capital,
- reducing costs,
- increasing the adaptability of new products,
- increasing of total profits and
- timely delivery of products to a well- known buyer.

Table 6.4 Lean and traditional production

Machinery/equipment	Expensive, specialized s	Small and highly flexible
Goals	Win competition	Competes for customers
Culture	Solved problems	Prevent problems
Priorities	The results	The results and processes
Procedure	Static	Dynamic
Employees	The cost of and troubles	Potential and opportunities
Machinery/equipment	Expensive, specialized s	Small and highly flexible
Resolving	"Who's to blame?", crisis	"What is the solution?", a source of improvement

The comparison of lean and traditional production is shown on Table 6.3.

Lean production can be summarized as a set of individual entities, principles and measures, which provide an effective consolidated form of unbroken chain in the creation of new shares. The overall concept is planned and managed through phases so a commercial-production system can reach its goal. It is necessary to ensure connectivity and eligibility of all employees and elimination of old, inappropriate organizational structures.

Applying the principles of lean production lead, in existing production systems, to:

➢ shortening the cycle of production,
➢ reduction of capital and
➢ reduction of required number of workers.

For the purpose of reducing the number of pieces in the series, it is necessary to leave the old divisions of labour and more convenient to introduce a system of team work, team responsibilities and team payment (instead of individual pay-per-action, which is today the most applicable).

The most important differences with respect to the previous mode:

➢ the more subcontractors than own production,
➢ inclusion in the sub-phases of the product and production, appropriate,
➢ not the principle of complete JIT (just for parts A – ABC analysis) and
➢ instead of self-control.

The most important instruments of lean production are:

➢ development and construction according to the demands of development, assembly, recycling and control,

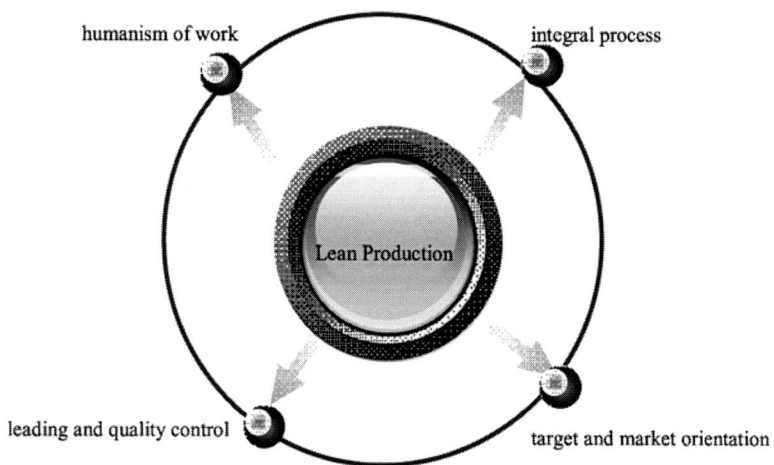

6.17 Segments of lean production.

> ➤ producers fully responsible for time and quality of product,
> ➤ full compliance deadlines for all associates,
> ➤ uninterrupted flow of material with minimum warehouses and stockpiles,
> ➤ timely decisions (make or buy),
> ➤ timely information from the market and
> ➤ correct assessment of the capacities of their own production.

The segments of lean production are shown on Figure 6.17.

The logistic chain of supplier-produce-consumer must function properly, which means that all factors in the chain should:

- keep the same policy of business-production system,
- establish relationships that lead to a joint goal and
- remedy any possible conflicts and obstacles.

Basic elements of Lean production (Figure 6.18) are in fact the Japanese methods and systems: JIT, Kanban, TQM, TPM, Kaizen, 5S, Six Sigma, 20 keys, etc.

6.3.3.1 *Techniques and tools of lean production*

Techniques and tools of lean production are: Six Sigma, 5S, JIT, Kaizen, Kanban, Error Proofing, Current Reality Trees, Conflict Resolution Diagram, Future Reality Diagram, Lean Metric, SMED (Single Minute Exchange of Dies), Standard Work, Takt Time, Total Productive Maintenance (TPM), Value Stream Mapping, Workflow Diagram, ect.

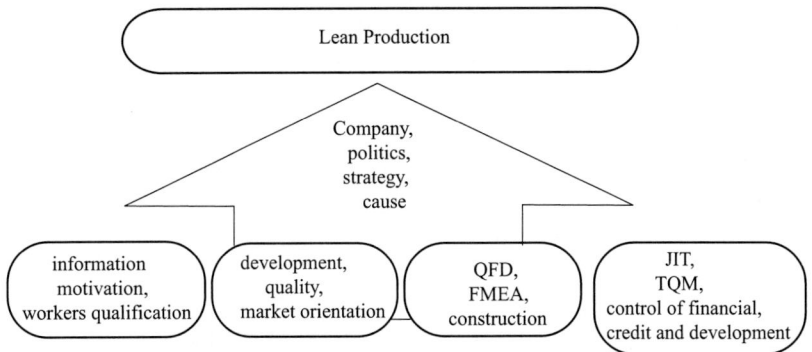

6.18   Basic element of lean production.

Six Sigma methodology is a set of methodological principles and statistical tools, which together give excellent results. The methodology was developed in Motorola, in the mid-eighties of last century, and was developed for the analysis of the production process and for eliminating errors. Effectiveness and efficiency of this methodology has been proved in many leading companies (General Electric and Texas Instruments).

The name Six Sigma is a statistical concept, which refers to six standard deviations. In statistical terms Six Sigma means 3.4 defects per million opportunities (DMPO), where sigma represents the variation in relation to the average value of the process.

In business terms Six Sigma is defined as a business strategy that is used to improve business profitability by eliminating errors, reducing costs of poor quality and improving the effectiveness and efficiency of all operations in order to meet or exceed the needs and expectations of customers.

Six Sigma methodology combines tools for continuous process improvement. The processes are analyzed and resources are objectively assigned to those processes that require the most attention. Errors in the process cause processing, spoilage, additional work, increased costs, etc. Focusing on prevention of errors and their efficient and effective remedy will reduce the labour standards and costs of processes, so resources can be released for other investments, and by comparing processes objective decisions about where to deploy resources can be made.

Two main Six Sigma methodologies are applied: DMAIC (Define, Measure, Analyze, Improve, Control) and DMADV (Define, Measure, Analyze, Design, Verify).

DMAIC method is used to improve a business process, and consists of five phases:

(1) Define – first it is necessary to define the project goal and purpose of the project and gather information about the process. At this stage the

following tools are applied: Affinity Diagram, Communications Plan, Control Charts, CTQ (Critical to Quality), Conventional Data, Kano Model, Pareto chart and SIPOC (Suppliers, Inputs, Process, Outputs, Customers) diagram.

(2) Measure – measure the existing situation with different tools: Control diagrams, Conventional Data, Flowchart, Histogram, Measurement system analysis (MSA), Defining Operations, Pareto chart, Six Sigma Analysis and Taguchi Loss Function.

(3) Analyze – analyze the causes using different tools: Brainstorming, Ishikawa diagram, DOE (Design of Experiments), Histogram, Testing Hypotheses, Diagram, Control diagram and Tree diagram.

(4) Improve – various improvements of processes or their parts using different tools: Network diagram (Gantt chart), Brainstorming, Control diagrams, Failure mode and effects analysis (FMEA), Histogram, Pareto chart, PDCA cycle and Priority Matrix.

(5) Control – control of the process itself and its outputs. This phase uses the following tools: Communications Plan, Control Charts and PDCA cycle.

DMADV method is used when it is necessary to implement the process, to design a new one or to restructure PBS. The very methodology is similar to DMAIC method. Expert team for improving the process first defines a critical level of quality. Then the optimal quality product or process is designed.

5S is a system to reduce waste and optimize productivity through maintaining an orderly workplace and using visual cues to achieve more consistent operational results. Implementation of this method "cleans up" and organizes the workplace basically in its existing configuration, and it is typically the first lean method which organizations implement.

5S, abbreviated from the Japanese, are simple but effective methods to organize the workplace. The 5S's are:

Phase 1 –**Seiri** – Sorting: Going through all the tools, materials, etc., in the plant and work area and keeping only essential items. Everything else is stored or discarded.

Phase 2 - **Seiton** – Straighten or Set in Order: Focuses on efficiency. When we translate this to "Straighten or Set in Order", it sounds like more sorting or sweeping, but the intent is to arrange the tools, equipment and parts in a manner that promotes work flow. For example, tools and equipment should be kept where they will be used (i.e. straighten the flow path), and the process should be set in an order that maximizes efficiency.

Phase 3 - **Seisō** – Sweeping or Shining: Systematic Cleaning or the need to keep the workplace clean as well as neat. At the end of each shift, the work area is cleaned up and everything is restored to its place. This makes it easy to know what goes where and have confidence that everything is where it should be. The key point is that maintaining cleanliness should be part of the daily work – not an occasional activity initiated when things get too messy.

Phase 4 - **Seiketsu** – Standardizing: Standardized work practices or operating in a consistent and standardized fashion. Everyone knows exactly what his or her responsibilities are to keep above 3S's.

Phase 5 - **Shitsuke** – Sustaining: Refers to maintaining and reviewing standards. Once the previous 4S's have been established, they become the new way to operate. Maintain the focus on this new way of operating, and do not allow a gradual decline back to the old ways of operating. However, when an issue arises such as a suggested improvement, a new way of working, a new tool or a new output requirement, then a review of the first 4S's is appropriate.

A sixth phase "Safety" is sometimes added.

This method emerged as the 3S, and it spreads in the latest trends up to 7S. According to some authors this is a separate method, while the others treat it as an integral part of the Lean approach.

Kaizen (continuous improvement) in the Japanese management practices is a continuation of incremental improvements and improvements in quality, technology, processes, company culture, productivity, security and governance. This method involves all employees. The basic philosophy of Kaizen management is fast, simple and easy, but constant improving of operating efficiency. Kaizen requires small financial investments, but major changes in its views, the way of work and thinking of all employees.

Kaizen method is often found in other methods and techniques (JIT, Kanban, 5S, 20 keys).

There are two levels of kaizen: system or flow kaizen focuses on the overall value stream and process kaizen focuses on individual processes.

Value Stream is all of the actions, both value-creating and nonvalue-creating, required to bring a product from concept to launch and from order to delivery. These include actions to process information from the customer and actions to transform the product on its way to the customer

6.3.3.2   Case study

In Serbian garment companies lean production is not in use. In Serbia the reorganization of few garment companies have just started ("natural" work flow of manufacturing, control tact time, training of the employees, using trolley for transport between sewing machines). But new methods for rationalizing the system of manufacturing garment are required. That's why the analysis in a domestic company for production of men's shirts was made by trying to implement lean production systems.

Analyzing production in the company for producing men's shirts discovered several causes of bad organization such as:

(1)   Causes from the work areas of employees

- poor organization of work,
- poor transport material,
- insufficient training of workers,
- poor working conditions,
- weak protection in the workplace,
- fluctuation of employees,
- inadequate schedule of work and rest,
- inappropriate system of compensation,
- poor interpersonal relations and
- various subjective reasons.

(2)   Causes in the field of using automation and funds:

- poorly organized maintenance service and repair of sewing machines,
- unprofessional and irresponsible management of production lines,
- chain system of installing jobs with immovable tapes,
- insufficient knowledge about the features of machines and devices,
- insufficient usage of capacities and
- bad choice of machines by capacity and type.

(3)   Causes in the area of textile materials and energy:

- accumulation of materials in the warehouse of raw materials,
- bad utilization of textile materials (big waste),
- insufficient control,
- irrational use of waste (for children's clothes),
- bad application of the basic textile materials and support,
- application of inappropriate machines and devices,
- errors in the construction of clothing,
- bad schedule and the location of energy sources and
- bad installations.

(4)   Causes due to methods, i.e. organizational procedures:

- insufficient preparation of the production of clothing,
- lack of coordination within the company,
- poorly organized records and control
- lack of work discipline

6.19 Inadequate storage of textile materials: a) plant b) cutting room.

- ■ protection at work
- ■ ignorance in the field of ergonomics and
- ■ incomplete technical and technological documentation.

Before the reorganization of the company for the production of men's shirts, interviewing 26 workers in a production line was carried out and the following results were obtained by method of 5S:

(1)   The research on sorting shows that 53.8% of workers believe that there is a large number of unnecessary machinery and equipment in the facility, while 98% of workers point out that textile materials are poorly housed partly on the plant floor, and partly in the appropriate shelves and rarely in the warehouse of raw materials (Figure 6.19). Despite damaging the quality of materials, it takes free space for transport.

(2)   The research on systematization shows that employees know where supplies and tools for work are and to return them regularly on their place, but there is not enough room for manipulation between workplaces, which points out the need for planning new installation of workplaces, Figure 6.20.

(3)   Preparation of a plant and the establishment of order and cleanliness of the premises as well as tools and equipment is satisfactory in the opinion of all respondents.

(4)   The tendency towards "zero defects" is not present among the workers, because only 15.3% of workers points out their and other people's mistakes or worry about the garment quality. Others who were questioned consider that the errors should be taken care of by the supervisor at the final control of products but believe that all the employees are responsible for the implementation of production.

(5)   After going into the 5S model, 57.6% of workers would accept it, while 19% oppose to any changes (most of the workers with over 25 years of service).

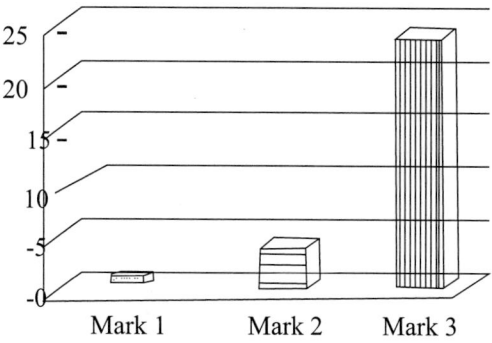

6.20 Rating adequate spaces for wheelchair transport.

6.21 Sewing machines that do not participate in the technological process.

On the basis of the survey for the possibility of applying 5S, the following conclusions can be made:

– It is necessary to insert additional racks for adequate disposal of textile materials in the production plant.
– It is necessary to move from the production line those sewing machines that do not participate in the technological process (mostly inoperative, Figure 6.21), or design workplaces so flexibility is achieved by technological trolleys, regardless of the model or item that is produced, as well as a required handling space.

- It is necessary to maintain sewing machines constantly.
- Provide the equipment and auxiliary devices that will accelerate and simplify technological operations.
- It is necessary to establish more self-control than control (team responsibility).
- It is necessary to train workers constantly so that they should accept the necessary changes.
- Provide a continuous flow of material with minimal storage and inventory.

No matter whether the work is being done for the unknown or a known customer (German "Lohn" work or Cut Make Trim system), the aim is to achieve a shorter manufacture time. The garment manufacturers in Serbia work more than 80% by CMT (Cut-Make-Trim) system, although most managers know that CTM jobs have no future in the world market. Their wages are getting lower and the competition is getting larger, so they are at risk of losing clients by any raising of prices. OBM (Original brand name manufacturing) system in the production for international markets is most likely currently unattainable for everyone but the majority of manufacturing firms have a competitive apparel.

It is therefore necessary to define a flexible model of production of clothing that will shorten the time of the technological process of cutting, sawing, and finishing. Application of CAD system accelerates the construction preparation (modelling, completing, duplicating the required number of sizes and the production of cutting layer), and thereby shortens the time of shirt production. Technological process of sawing can be shortened only by investing in automated machine for laying textile material, or into a modern CAM system for depositing materials and cutting parts. In Tables 6.5,

Table 6.5 Analysis of technological cutting process

Technological process	Construction preparation (s)	Technological cutting process		
		Placing of material (s)	Cutting (s)	Thermo fixation (s)
Traditional production	10 440	108	401	59
Lean production with CAD/CAM	2880	50	200	59
Lean production without CAD/CAM	10 440	108	401	59

Table 6.6 Analysis of technological sewing process

| Technological process | Technological sewing process (s) | | | | | | | |
	Making collars	Making pockets	Making right front parts	Making left front parts	Making back	Making sleeves	Making cuffs	Montage
Traditional production	207	53	94	94	177	75	160	154
Lean production with CAD/CAM	207	**25**	94	**69**	177	75	160	154
Lean production without CAD/CAM	207	**25**	94	**69**	177	75	160	154

Table 6.7 Analysis of technological finishing process

| Technological process | Technological finishing process | | |
	Final ironing (s)	Quality control and packing (s)	Warehouse of ready-made garment (s)
Traditional production	108	401	59
Lean production with CAD/CAM	50	200	59
Lean production without CAD/CAM	108	401	59

6.6 and 6.7 the times of technological processes of cutting, sewing and finishing are shown.

Technological process of making men's shirts consists of 60% of manual work and 40% of machine work. By the application of the study of work the resulting model can be further improved by the reduction of time, primarily of manual work in a sewing room.

In the work on the sewing machine and the automats there is still a large percentage of manual work due to poor shaping of workplace and internal

Table 6.8 Average production time and making men's shirts and suggested models

	Production time		Time for manufacturing men's shirts	
	s	min	s	min
Traditional production	12 649	210	1939	32
Lean production with CAD/CAM	4777	80	1627	27
Lean production without CAD/CAM	12 596	209	1886	31

transport. By detailed analysis of all jobs in the production line, the time of making a garment can be reduced for a few more minutes per unit of product.

In the Table 6.8 is shown that the technological process of sewing and finishing a production line can be reduced only for 1 min without the application of CAD/CAM systems.

New model of production men's shirts can be checked through the daily capacity of C_d. In one production line (sawing and finishing) there are the average of 20 workers with work time 7.5 h (450 min).

According to that, the daily capacity is:

- C_d traditional model = 391 pieces
- C_d research model = 409 pieces

Annual capacity (C_g) of a model of a production line (240 working days) and cost effectiveness of models are shown in Table 6.9.

Percentage of savings for the production of male shirt with the 99.160 pieces in relation to the average production with 93.840 pieces is 4.6 % and that increases revenue production lines a year for 69 120 euros per year.

The analysis of production in company for men's shirts demonstrates us one way for change organization in our garment manufacturing, because:

o reflects and supports target attainment and quality values for short term and long-term periods,

Table 6.9 Year capacity (Cg) and increase revenue

	Year capacity, piece	Price (16 € for 1 shirt)
Traditional production	93 840	1 501 440
New production	98 160	1 570 560
Between	4320	**69 120**

- development and engagement of all employees for improvement within the organization,
- resources of an organization (finance, IT, height-tech textile material and new cutting and sewing technologies) are coordinated with the quality of garment and organization values,
- overview of all processes in a garment company and change of the existing combination of processes, emphasis on shortening the technological time,
- indirect connection with customer satisfaction,
- organization will be successful only if it adequately motivates its employees,
- quantitative evaluation as better quality, increase of productivity and
- reduced stocks.

In order to make the technological system of production flexible, it is necessary to:

- Be always on the training.
- New practice of knowledge and experience in the field of management.
- New knowledge about practice in the field of garment technology and information systems.
- Set of positions to achieve a faster transport of objects and to answer any of the technological procedures (different models and articles).
- The constant practice of training workers.
- Work on the acceptance of change.
- Analyze the technological operations and procedures and make their optimization.
- Design workplace.
- Improvement of internal transport.
- Apply the techniques of network planning of production.
- Define each of the quality of fashion products.
- The trend towards the "zero error" and reduce the warehouse of finished products.
- Bringing the team responsibility of all employees.
- The creation of recognizable fashion brands.

References

1. Bagozzi R P (1994), *Principles of Marketing Research*, Blackwell Business, Cambridge, MA, USA.
2. Barnard W and Wallace T F (1994), 'The Innovation Edge: Creating Strategic Breakthrough Using the Voice of the Customer', *Oliver Wight Publications*, New York

3.   Berg P, Appelbaum E, Bailey T and Kalleberg A L (1994), 'The Performance Effects of Modular Production in the Apparel Industry', *Industrial Relations*

4.   Benjaafar S and Ramakrishnan R (1995), 'Modelling, measurement and evaluation of sequencing flexibility in manufacturing systems', *The International Journal of Production Research*

5.   Bolwijn P (1986), *Flexible manufacturing integrating technological and social innovation,* Elsevier, Amsterdam, New York

6.   Bolwijn P and Kumpe T (1990), 'Manufacturing in the 1990's- Productivity, flexibility and innovation', *Long range planning*, 23(4)

7.   Broyles D, Franko J and Bergman M (2005), *Just In Time Inventory Management Strategy Information*, Kansas State University

8.   Nilsson C H and Nordahl H (1995), 'Making manufacturing flexibility operational – part 1: a framework', *Integrated Manufacturing Systems*, 6 (1)

9.   Colovic G. Paunovic D and Savanovic G (2008), 'Primena prosirene pametne organizacije proizvodnje odece', *11. Medjunarodna konferencija ICDQM-2007, Upravljanje kvalitetom i pouzdanoscu*, Beograd, 890–893

10. Colovic G, Paunovic D and Savanovic G (2009), 'The Analysis of Some Quality Methods in Garment Industry', *8th International Scientific Conference of Production Engineering, RIM 2009*. Kairo

11. Colovic G, Paunovic D and Savanovic G (2009), 'Modelling of Flexible Technological of Garment Production Process By Using Modern Information Technology', *XIV Научна конференция с международно участие ЕМФ 2009,* Созопол

12. Colovic G (2009), 'Lean Production in the Serbian Garment Industry', *12th International Conference "Dependability and Quality Management", ICDQM-2009*, p5-43, plenary lectures

13. Colovic G, Paunovic D and Paunovic S (2009), 'Značaj OBM za razvoj odevne industrije Srbije', *V Majska konferencija o strategijskom menadzentu,* Bor, pp 127 -131

14. Crevelling C.M (2002), *Design for Six Sigma in Technology and Product Development*, Prentice Hall, New Jersey

15. Crosby D and Kimber R (2008), *The Zero Defects Option*, The Crosby Company

16. Dean E B (1996), *Quality Function Deployment for Large System*, NASA Langley Research Center, Hampton

17. Deming W E (1996), *Nova ekonomska nauka*, Grmec, Beograd

18. Djordjevic D and Djekić I (2001), *Osnove upravljanja kvalitetom*, Teagraf, Beograd

19. Djuricic M, Paunovic D and Djuricic M.M (2008), 'Application of the QFD method in clothing industry', *International Conference Science And Higher Education In Function Of Sustainable Development, SED 2008*, Uzice

20. Dooner M (1991), 'Conceptual modelling of manufacturing flexibility', *International Journal of Computer Integrated Manufacturing*, Vol. 4, No. 3, 135–144

22. Dunlop T and Weil D (1996), *Diffusion and Performance of Modular Production in the U.S. Apparel Industry*, Harvard Centre for Textile and Apparel Research, Industrial Relations

23. Gaither N (1992), *Production and Operations Management*, The Dryden Press, International ed.

24. Garg M (2003), 'Moving From Lean or Six Sigma to Lean and Six Sigma', *Operations Management*, Issue No. 2., SPJIMR Bullet

25. Hunter A (1990), *Quick Response in Apparel Manufacturing: A Survey of the American Scene*, The Textile Institute

26. Ishikava K (1990), *Introduction to Quality Control*, J. H. Loftus, 3A Corporation, Tokyo
27. Ishikawa K (1985), *How to Operate QC Circle Activities*, QC Circle Headquarters, Union of Japanese Scientists and Engineers, Tokyo
28. Juran J M (1997), *Oblikovanjem do kvaliteta*, Grmec, Beograd
29. King R and Hodgson T (2001), *Analysis of Apparel Production Systems to Support Quick Response Replenishment*, National Textile Centre Annual Report
30. Kobayashi I (1995), *20 Keys to Workplace Improvement*, Productivity Press, Portland
31. Lazic M (2005), *Alati, metode i tehnike unaprednja kvaliteta*, Masinski fakultet, Kragujevac
32. Lowson B, King R and Hunter A (1999), *Quick Response*, Wiley, New York
33. Martinovic M and Colovic G (2007), 'System PPORF in Garment Industry', *Serbian Journal of Management*, 2 (1), 73–81
34. Monden Y (1983), *Toyota Production System*, Industrial Engineering and Management Press, Norcross
35. Monden Y (1998), *Toyota Production System, an Integrated Approach to Just-In-Time*, Third edition, Norcross, GA: Engineering & Management Press
36. Nagarkar S and Bennett D J (1988), '*Flexible manufacturing system lets small manufacturer of mainframes compete with giants*', *IE*, 20 (11)
37. Paunovic D, Colovic G and Nikolic V (2009), 'The quality function deployment method in garment industry', *Communications in Dependability and Quality Management*, vol. 12, (2)
38. Paunovic D, Djuricic M and Savanovic G (2008), 'FMEA Analysis in garment industry', *International Scientific Conference*, Gabrovo
39. Russell R and Taylor B (1998), *Operations Management: Focusing on Quality and Competitiveness*, Prentice Hall, New Jersey
40. Shingo S (1988), *Non-stock Production, The Shingo System for Continouns Improvement*, Productivity Press, New York
41. Shingo S (1989), *Study of the Toyota Production System*, Productivity Press, New York
42. Tague N (2005), *The Quality Toolbox*, Second Edition, ASQ Quality Press, Milwaukee
43. Zelenovic D (1986), *Upravljanje proizvodnim sistemima*, Naucna knjiga, Beograd
44. Womack J, Roos D and Jones D (2003), The Machine That Changed The World: The Story of Lean Production, TPM

Index